SpringerBriefs in Molecular Science

For further volumes:
http://www.springer.com/series/8898

Yao He · Yuanyuan Su

Silicon Nano-biotechnology

Yao He
Yuanyuan Su
Institute of Functional Nano and Soft
  Materials (FUNSOM),
  Jiangsu Key Laboratory for Carbon-
  Based Functional Materials & Devices
  and Collaborative Innovation Center of
  Suzhou Nano Science and Technology
Soochow University
Suzhou
China

ISSN 2191-5407   ISSN 2191-5415   (electronic)
ISBN 978-3-642-54667-9   ISBN 978-3-642-54668-6   (eBook)
DOI 10.1007/978-3-642-54668-6
Springer Heidelberg New York Dordrecht London

Library of Congress Control Number: 2014934458

Printed on acid-free paper

Springer is part of Springer Science+Business Media (www.springer.com)

Preface

In the past decade, we have witnessed the giant advancement of silicon nanotechnology, which provides exciting new avenues for myriad electronic, energetic, environmental, biological, and biomedical applications. Among them, the exploration of silicon nanotechnology for bioapplications (so-called silicon nano-biotechnology) is one of the most important branches, receiving extensive attentions and revolutionizing basic research and clinical applications in recent years. Therefore, based on the previous elegant work of scientists worldwide and recent progress of our group, we publish this book that introduces silicon nanotechnology for biological and biomedical applications, particularly for biosensing, bioimaging, and cancer therapy. It is worthwhile to point out that, compared to the sufficiently published reports, only limited references are cited here due to the page limitation. Therefore, we express our apologies to all the scientists whose research work is not introduced in this book. The present book may potentially serve as a new starting point in the realm of silicon nano-biotechnology, and will be of interest to all chemists, material scientists, as well as biologists and clinicians.

We express our sincere thanks to Prof. Shuit-Tong Lee for his generous help and invaluable suggestions. We are thankful to Mr. Fei Peng (a Ph.D. student under Prof. Yao He's supervision) for his kind help in the elaborate and systematic literature investigation. We appreciate the financial support from the National Basic Research Program of China (973 Program 2013CB934400 and 2012CB932400), the Funds for International Cooperation and Exchange of the National Natural Science Foundation of China (Grant No. 61361160412), the Natural Science Foundation of Jiangsu Province of China (Grant No. BK20130052 and BK20130298), the Specialized Research Fund for the Doctoral Program of Higher Education of China (Grant No. 20133201110019 and 20133201120024), and a project funded by the Priority Academic Program Development of Jiangsu Higher Education Institutions (PAPD).

Yao He
Yuanyuan Su

Contents

Chapter 1
Introduction

Nanotechnology has been widely regarded as one of the most important break-throughs since the last century, significantly revolutionizing science and technology in the past several decades. As officially defined by the US National Nanotechnology Initiative in 2000, "Nanotechnology is concerned with materials and systems whose structures and components exhibits novel and significantly improved physical, chemical and biological properties, phenomena and processes due to their nanoscale size" [1]. Materials with at least one dimension sized from 1 to 100 nanometers (so-called nanomaterials) generally exhibit new and unique optical/electronic/magnetic merits, serving as essential and important tools for nanotechnology applications [2]. Thus far, various kinds of functional nanomaterials (e.g., semiconductor nanomaterials, carbon nanomaterials, and silicon nanomaterials, etc.) have been well developed [1, 3–5], which offers exciting opportunities in virtually all branches of nanotechnology ranging from optical systems, electronic, chemical, and automotive industries, to environment, engineering, biology, and medicine. Among them, nanotechnology for biological and medical applications (generally described as "nano-biotechnology") is considered as one of the most important braches, which has shown great promise from basic research (e.g., investigation of complicated biological and biomedical processes that are hard to access with conventional approaches) to practical applications (e.g., early diagnosis and treatment of diseases) [5, 6].

Silicon is well-known as the crust's second most abundant element on earth, only behind oxygen, providing a rich and low-cost resource support for myriad silicon-based applications. By virtue of its excellent semiconductor and mechanical properties, silicon materials act as the leading semiconductor materials and dominate the electronics industry to date. Notably, novel structural, optical or/and electronic characters emerge when the dimensions of silicon materials are reduced to nanoscale level (so-called silicon nanomaterials) [7–9]. The last 20 years have witnessed the vast advancement in fabricating silicon nanomaterials and the rapid development of silicon nanomaterials-based applications in various fields, including electronics, energy, environment, biology, and biomedicine [10–12]. Taking advantage of non- or lowly toxic property of silicon, exploration of silicon nanotechnology for biological and biomedical applications is of particular interest

Y. He and Y. Su, *Silicon Nano-biotechnology*, SpringerBriefs in Molecular Science, DOI: 10.1007/978-3-642-54668-6_1, © The Author(s) 2014

Fig. 1.1 Silicon
nanobiotechnology holds
great promise for myriad
biological and biomedical
applications, particularly for
biosensor, bioimaging, and
cancer therapy, etc

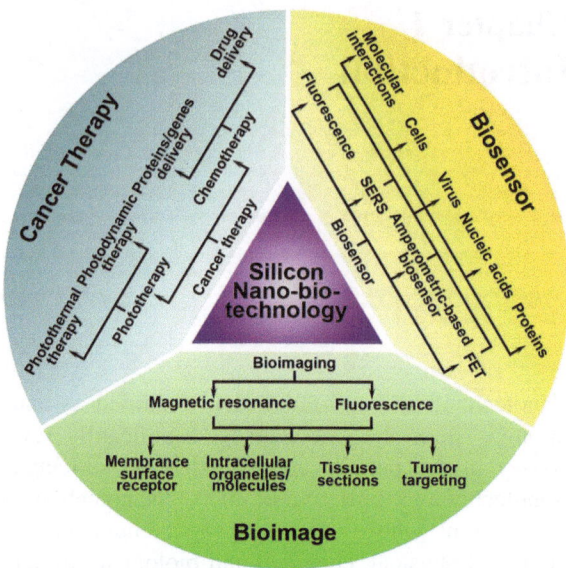

and has been extensively studied in recent years. This chapter presents an introduction of silicon nanomaterials, followed by a brief summary of silicon nanotechnology for various bioapplications, particularly for biosensor, bioimaging,
cancer diagnosis and therapy, etc. (Fig. 1.1).

1.1 Fabrication of Silicon Nanostructures

Scientists have made an elegant work to successfully fabricate a great number of
nanomaterials with well-defined structures and required functionality. To date,
metal (e.g., silver and gold) nanostructures (e.g., nanoparticles, nanorods, and
nanoshells) [13–15], fluorescent semiconductor II–VI quantum dots (QDs) [16–19],
carbon-based nanostructures (e.g., carbon nanotubes and graphene) [6, 20, 21],
magnetic nanoparticles (e.g., Fe_3O_4 nanoparticles) [22–24], and silicon nanostructures (e.g., silicon nanoparticles and nanowires) [7, 8, 25] have been well developed.

Silicon nanoparticles (SiNPs) and silicon nanowires (SiNWs) are recognized as
the most important zero- and one-dimensional silicon nanostructures, respectively.
Of particular note, SiNPs whose sizes are generally smaller than ~5 nm, exhibit
relatively strong fluorescence due to effective recombination of electron and hole
confined in nanodots, leading to a promising perspective of long-awaited optical
applications [26, 27]. To date, several synthetic approaches have been well
established for preparation of fluorescent SiNPs. Typically, in the 1990s, Kauzlarich and coworkers introduced a solution-phase reduction synthesis strategy,
which is capable of mildly producing SiNPs at room temperature and normal

atmospheric pressure [28]. Kortshagen et al. introduced a method titled plasma-assisted aerosol precipitation for high-yield synthesis of SiNPs with controllable sizes ranging from 2 to 8 nm [29]. In 2009, Lee et al. reported a electrochemical etching method, allowing fabrication of multicolor fluorescent SiNPs with tunable maximum emission wavelengths from 450 to 740 nm [30]. It is worthwhile to point out that, most of the above-mentioned SiNPs generally possess poor aqueous dispersibility, thus requiring additional post-treatment or surface modification to render the prepared SiNPs hydrophilic for biological and biomedical applications. To address this issue, He et al. recently developed novel microwave-assisted strategies to facilely and directly synthesize highly fluorescent and water-dispersed SiNPs in aqueous phase [31–33]. In particular, one of their latest studies demonstrated that the large-scale SiNPs with excellent aqueous dispersibility, strong fluorescence, and robust stability could be rapidly achieved (e.g., 0.1 SiNPs were yielded in 15-min microwave reaction) [33].

On the other hand, a variety of methods (e.g., chemical vapor deposition (CVD) [34–38], oxide-assisted growth (OAG) [25, 39], and metal-catalyzed electroless etching [40–42], etc.) have been well established for preparation of SiNWs. Specifically, due to elegant work of Lieber, Lee, Yang, et al., CVD and OAG [25, 37–39] have been widely employed as two most popular means to fabricate SiNWs and SiNWs arrays with high aspect ratio and production yield. Recently, Peng, Lee, et al. developed a class of metal-catalyzed electroless etching approach (e.g., HF-etching-assisted nanoelectrochemical method) [40–42], serving as an alternative method to facilely produce SiNWs in a low-cost manner. In addition to SiNPs and SiNWs, silicon-based nanohybrids featuring multifunctional properties are promising as powerful tools for various applications [43]. Figure 1.2 presents typical photographs and microscopical images of SiNPs, SiNWs, and SiNWs-based nanohybrids. Sufficient details concerning the design of SiNPs, SiNWs, and their nanohybrids will be discussed in Chap. 2.

1.2 Biosensor

Analysis and detection of chemical and biological species is of essential importance for biomedical diagnosis, food safety, environment monitoring, and anti-bioterrorism, etc. [44]. A high-performance biosensor is expected to be highly sensitive, specific, reproducible, biocompatible, and capable of simultaneously multiple target detection. In principle, two modules, i.e., a recognition element for target binding and a transduction element for signaling the binding event are basically included in a sensor. Nowadays, a great deal of sensing devices can be commercially purchased. However, novel kinds of biosensors with sufficient sensitivity, high specificity, excellent reproducibility, and multiplexing detection capabilities still remain in high demand. Nanomaterials featuring unique physicochemical properties, serving as novel recognition or transduction elements, provide new possibilities for designing high-quality bioassay kits [45]. In

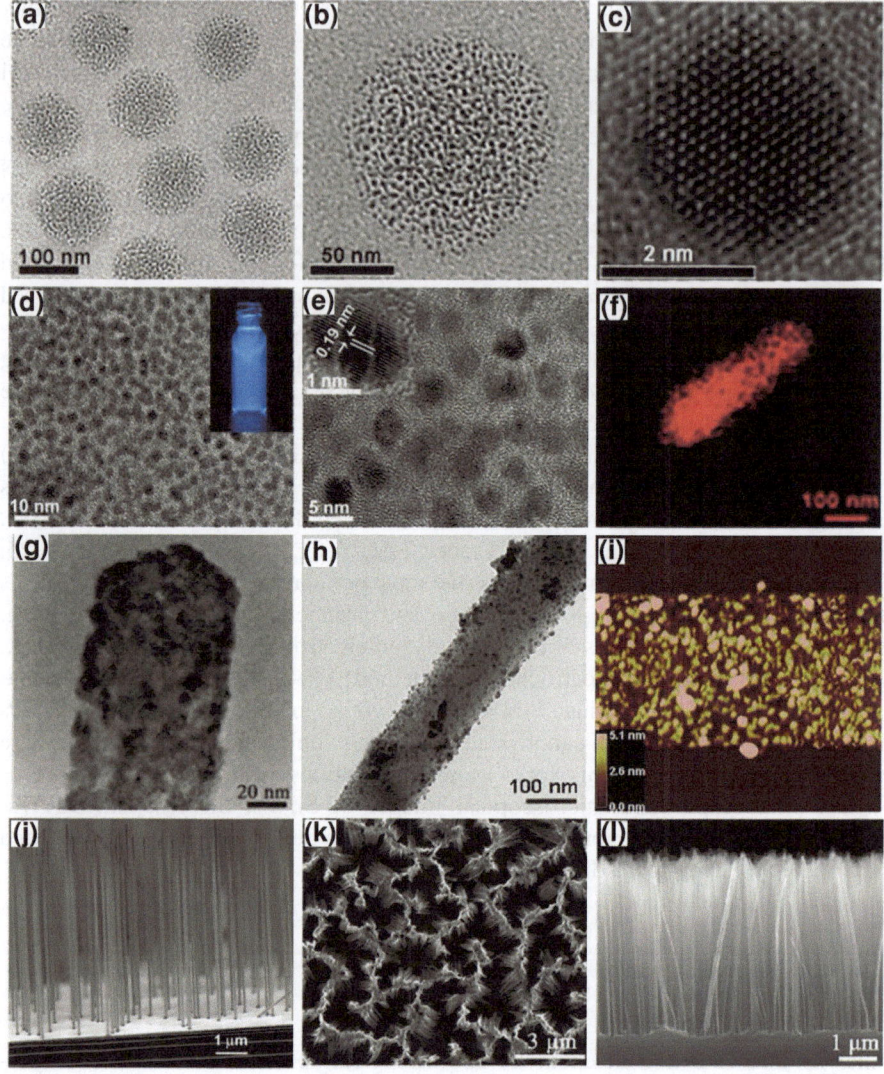

◄ **Fig. 1.2** Photographs and microscopical images of typically silicon nanostructures. **a** and **b** Transition electron microscopy (TEM) images of the polymer-modified silicon nanospheres (SiNSs) with a diameter of ~120 nm. **c** The high-resolution TEM image of a single silicon nanoparticle inside the as-prepared nanospheres. Reproduced from Ref. [79] by permission of John Wiley & Sons Inc. **d** TEM and **e** HRTEM images of water-dispersed fluorescent silicon nanoparticles (SiNPs). Inset in (**d**) presents a picture of the SiNPs aqueous sample excited by UV lamp with a maximum emission wavelength of 365 nm. Inset in (**e**) presents the HRTEM image of a single SiNP. Reprinted with the permission from Ref. [33]. Copyright 2013 American Chemical Society. **f** and **g** Fluorescence and TEM images of the red-emitting quantum dots (QDs)-decorated SiNWs. Reproduced from Ref. [43] by permission of John Wiley & Sons Inc. **h** and **i** TEM and AFM images of a single silicon nanowire (SiNW) whose surface is coated by a large number of silver nanoparticles (AgNPs). Reprinted from Ref. [66]. Copyright 2011, with permission from Elsevier. **j** Scanning electron microscopy (SEM) cross-sectional image of the SiNW array. Reprinted with the permission from Ref. [35]. Copyright 2005 American Chemical Society. **k***Top* and **l** cross-section SEM images of SiNWs arrays prepared via metal-catalyzed electroless etching approach. Reprinted with the permission from Ref. [70]. Copyright 2012 American Chemical Society

particular, sensitivity of nanobiosensors can be remarkably ameliorated by utilizing large surface-to-volume ratio of nanomaterials. Besides, abundant surface chemistry and facile surface modification of nanomaterials vastly facilitate the improvement of sensing specificity. Moreover, large surface area of nanomaterials affords the possibility to design sensing devices for multiplexed detection. Consequently, multifarious nanobiosensors have been constructed by employing functional nanomaterials (e.g., silver/gold nanoparticles, carbon nanotubes, semiconductor quantum dots, and graphene, etc.) as novel sensing platform, allowing sensitive, specific, and multiplexed detection of nucleic acids, protein, cells, and so forth [46].

Taking advantage of the excellent electronic/mechanical properties, huge surface-to-volume rations, facile surface modification, and compatibility with well-developed silicon technology [10, 25], SiNWs are considered as promising candidates for fabrication of high-performance biosensors [47, 48]. As shown in Fig. 1.3, based on different types of detection signals, the SiNWs-based biosensors are dominantly categorized as electrochemical and optical sensing devices. Among electrochemical sensors, SiNWs-based field-effect transistor (FET) [49] is the most widely-studied one, which has been employed for detection of widespread biological targets, including DNA [50–53], proteins [54, 55], virus [56], as well as the interaction between biological species [57]. Moreover, SiNWs can be also utilized as novel electron transfer mediator for electrochemical sensing applications. For example, SiNWs with subtle electronic properties and huge surface-to-volume ratio are configured as a working electrode, facilitating the enhancement of electrochemical reactivity and promotion of electron transfer reactions of biomolecules [58, 59]. On the other hand, SiNWs, acting as sensing platform, are suitable for constructing optical biosensors. Notably, silver/gold nanoparticles (NPs)-decorated SiNWs have demonstrated to be highly active surface-enhanced Raman scattering (SERS) substrates with large enhancement factor (EF) values, which are generally up to 10^6–10^9 [60–63]. A consensus has been reached that,

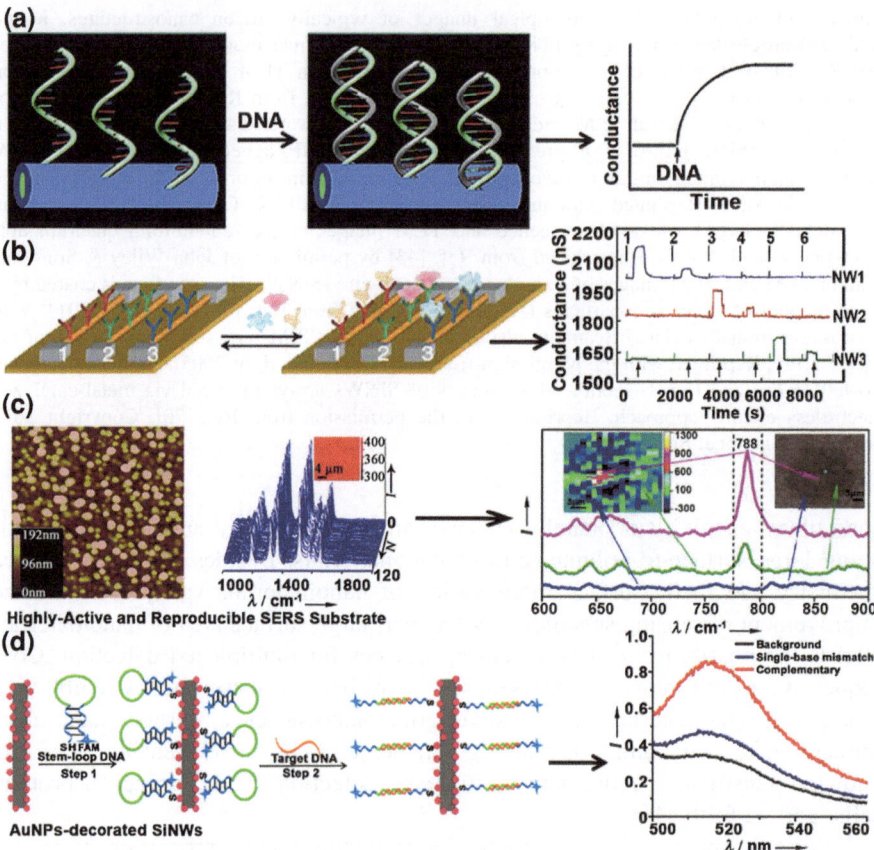

Fig. 1.3 Schematic diagram of silicon materials-based nanobiosensors. **a** Peptide nucleic acid (PNA)-modified SiNW-based sensors capable of DNA detection, based on observing conductance response from the resultant sensors in a real-time manner. Reprinted with the permission from Ref. [50]. Copyright 2004 American Chemical Society. **b** Schematic illustration of SiNW array-based sensing devices for multiple detection of various proteins. SiNW1, SiNW2, and SiNW3 are functionalized with three kinds of monoclonal antibody (mAbs) receptors, which are ready for simultaneous and specific detection of prostate specific antigen (PSA), carcinoma embryonic antigen (CEA) and mucin-1. Reprinted by permission from Nature Publishing Group, a division of Macmillan Publishers Ltd: Ref. [55], copyright 2005. **c** *Left*, atomic force microscopy (AFM) image of AgNPs-coated silicon wafer (AgNPs@Si) and Raman mapping spectra of R6G dispersed on the prepared AgNPs@Si surface (*left*). *Right*, representative SERS spectra and the corresponding SERS mapping image of one single A549 cell cultured on surface of the AgNPs@Si. Reprinted with the permission from Ref. [69]. Copyright 2013 American Chemical Society. **d** Schematic fabrication of SiNWs-based molecular beacons (MBs), allowing sensitive and specific detection of DNA. Reprinted with the permission from Ref. [70]. Copyright 2012 American Chemical Society

while free AgNPs/AuNPs are well-studied SERS substrates, they are nevertheless prone to be uncontrollably aggregated in aqueous phase, thus leading to poor reproducibility of SERS signals [64, 65]. In terms of metal NPs-decorated SiNWs, the NPs can be steadily immobilized on surface of SiNWs, efficiently preventing random NPs aggregation [63]. As a result, signal reproducibility of the SiNWs-based SERS substrates can be markedly improved. Moreover, AgNPs/AuNPs can be efficiently coupled through interconnection assisted by the semiconducting SiNW, yielding high-efficacy "hot spots" and giant SERS enhancement. Consequently, these SiNWs-based nanohybrids, serving as highly active SERS substrates, have been widely employed for construction of high-performance biosensors, allowing detection of a trace amount of biological species (e.g., nucleic acids and protein) with extremely low concentrations, even down to ~femtomole level [11, 60, 66, 67]. In addition to high sensitivity and reproducibility, the SiNWs-based SERS biosensors feature excellent multiplexed detection capabilities, allowing simultaneous detection of multiple DNA sequences. The reason is that the different kinds of DNA strands are readily assembled with SiNWs due to their large surface area and huge surface-to-volume ratios. For example, one recent report introduced a kind of SiNWs-based SERS sensors capable of detecting different tumor-suppressor genes (e.g., p21 and p53) [63]. Instead of SiNWs often involving relatively tedious synthetic procedures, silicon wafers are ready for direct in situ growth of AgNPs, facilely producing AgNPs-coated silicon wafers (AgNPs@Si). The resultant AgNPs@Si can be utilized for construction of high-performance SERS sensors with high sensitivity, specificity, reproducibility, and multiplexing capabilities, enabling detection of two types of DNA strands at ~pM level [68]. In the latest study, such silicon-based SERS substrates are further explored as in vitro sensing platform, which is suitable for analyzing apoptotic cells at the single-cell level [69]. In addition to SERS sensors, silicon nanomaterials have been used for design of fluorescence-based biosensors (e.g., SiNWs-based molecular beacons) to take advantage of their unique optical properties (e.g., high fluorescence quenching efficiency) [70]. In Chap. 3, we shall talk about representative achievements of silicon materials-based nanobiosensors in a detailed way.

1.3 Bioimaging

Bioimaging is one of the most significant imaging techniques that is superbly suited to direct visualization of biological systems, including labeling and tracking cells/animals, observing morphologies and behaviors of cells/tissues, and imaging subcellular organs (e.g., microtubules, nuclei, etc.), and so forth [18, 19]. Biological probes are essential tools for bioimaging applications. Functional nanomaterials with unique optical/magnetic merits offer exciting avenues for design of novel high-quality biological nanoprobes. For example, since the first report on II–VI semiconductor quantum dots (e.g., CdSe/ZnS QDs)-based fluorescent probes independently proposed by Alivisatos' and Nie's group in 1998 [16, 17], this kind of nanoprobes have been used for widespread bioimaging applications [18, 19, 71]. Compared to conventional fluorescent probes (e.g., fluorescent protein and organic dyes) with severe photobleaching property, the II–VI QDs-based nanoprobes of robust photostability are highly efficacious for long-term and real-time fluorescence bioimaging. As a result, biological targets or processes can be readily investigated in a long-term manner, which are difficult to access with conventional probes. Take magnetic nanoparticles for another typical example, Fe_3O_4 nanoparticles are currently regarded as highly promising candidates of contrast agents for magnetic resonance imaging (MRI) [22, 72, 73]. Compared to common MRI contrast agents (e.g., gadolinium) raising toxic gadolinium-induced safety concerns, magnetic Fe_3O_4 nanoparticles are much more biocompatible (e.g., biodegradation product of the Fe_3O_4 nanoparticles (i.e., iron ions) is a natural trace element existing in human body), and thus favorable for wide-ranging biomedical applications [72, 74]. Those exciting progresses indicate that nanotechnology would lead to new transformations in bioimaging technology. However, there still remain many major challenges in this area, such as heavy metals-induced safety concerns of II–VI QDs [75], unclear behavior of nanoprobes in vivo (e.g., biodegradability and organ accumulation), etc. [76].

Silicon is well-known as the trace element naturally existing in numerous tissues. Besides, recent studies reveal that porous SiNPs are biodegradable and can be readily cleared from a mice mode via renal clearance with feeble adverse effect in vivo [77]. These attractive merits have triggered intense investigation for exploring fluorescent SiNPs as novel biocompatible nanoprobes for bioimaging applications (Fig. 1.4). Thus far, several research groups have developed a number of SiNPs-based bioprobes for fluorescence bioimaging in vitro and in vivo [78–82]. Typically, hydrophilic molecules (e.g., acrylic acid and allylamine)-modified SiNPs and silicon nanospheres (SiNSs) containing hundreds of SiNPs have been primarily used for cell imaging [78–80]. However, meager pH stability of the above SiNPs and SiNSs (e.g., slight pH changes would result in severe fluorescent quenching of SiNPs) severely hampers their wide-ranging bioimaging applications, particular in immunofluorescent bioimaging. In comparison, micelle-encapsulated SiNPs-based probes preserve strong pH stability; nevertheless, the resultant probes often possess poor optical stability (e.g., quantum yield of SiNPs dramatically drops from ∼17 to 2 % due to micelle encapsulation), which is adverse to bioimaging sensitivity. Moreover, based on recent reports revealing nanomaterials with dynamic diameters smaller than 10 nm are relatively easily excluded from body without detectable toxicity, SiNSs and micelle-encapsulated SiNPs of large sizes (∼60–300 nm) are probably not favorable for biological applications [76, 83]. Very recently, He et al. successfully fabricate a novel class of SiNPs-based bioprobes featuring small size (∼4 nm) and strong fluorescence (quantum yield: 15–25 %) [33]. Of particular significance, this kind of silicon-based nanoprobes is superbly suited to immunofluorescent long-term cellular imaging due to their ultrahigh photo/pH stability, providing exciting opportunities for design of high-quality silicon nanoprobes and development of silicon-based

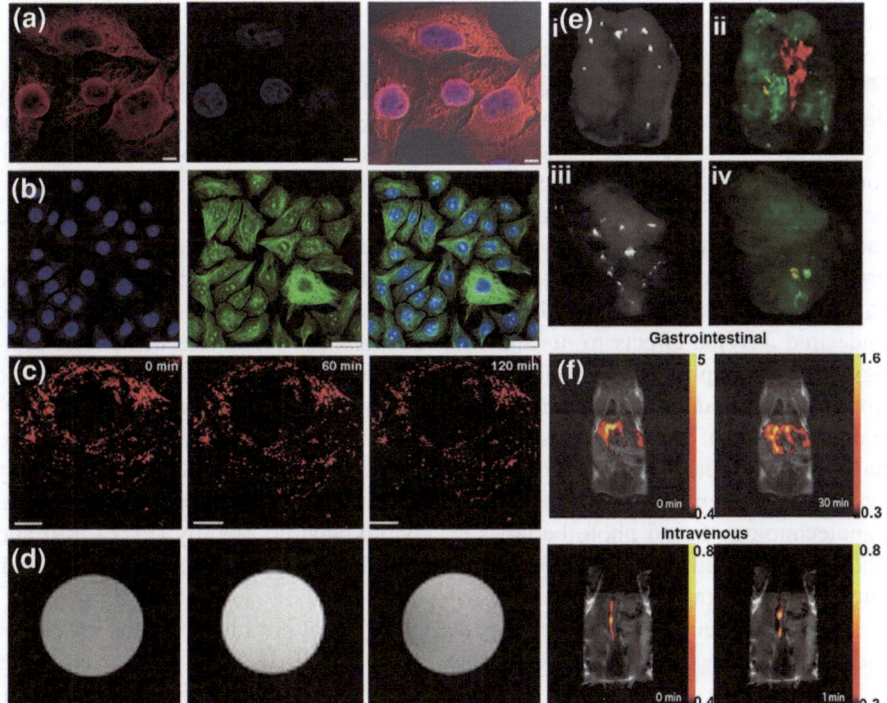

Fig. 1.4 a Double-color imaging of HeLa cells whose microtubules and nuclei are labeled by SiNWs-based nanoprobes (*red fluorescence*) and Hoechst (*blue fluorescence*), respectively. Reproduced from Ref. [43] by permission of John Wiley & Sons Inc. **b** Silicon nanoprobes-based Immunofluorescence imaging picture in which nuclei and microtubules are respectively targeted with SiNPs-based fluorescent probes (*blue*) and FITC (*green*) with high resolution. Reprinted with the permission from Ref. [33]. Copyright 2013 American Chemical Society. **c** Long-term cellular imaging using photostable SiNPs-based nanoprobes whose fluorescence is bright and stable during 120-min continuous observation. Reprinted with the permission from Ref. [31]. Copyright 2011 American Chemical Society. **d** T_1W Magnetic resonance images (MRI) of P388D1 macrophage cells labeled by the Mn-doped SiNPs (Si$_{Mn}$NPs) Reprinted with the permission from Ref. [87]. Copyright 2010 American Chemical Society. **e** Ex vivo (*i, iii*) and fluorescence images (*ii, iv*) of tumor tissue treated with (*i, ii*) micelle-encapsulated SiNPs conjugated with RGD peptides or or (*iii, iv*) pure MSiNP. Reprinted with the permission from Ref. [81]. Copyright 2011 American Chemical Society. **f** Hyperpolarized silicon particles-based ^{29}Si MRI of a mouse model in a real-time manner. Reprinted by permission from Nature Publishing Group, a division of Macmillan Publishers Ltd: Ref. [86], copyright 2013

fluorescence bioimaging techniques. On the other hand, multifunctional silicon nanomaterials (e.g., fluorescent and magnetic SiNPs) have recently been explored as novel probes for MRI applications, with a high contrast manner particularly suitable for in vivo imaging [84–87]. The above exciting achievement on development of silicon-based nanoprobes for bioimaging will be elaborately summarized and discussed in Chap. 4.

1.4 Cancer Therapy

Human beings all around the world have been facing the threat of cancer for many decades. Cancer, recognized as the most dangerous disease, kills millions of people each year. Scientists and doctors have made tremendous efforts to search for effective methods for cancer therapy; however, limited progress has been achieved thus far. While traditional therapeutic strategies (e.g., chemotherapy and radiotherapy) have been widely employed for the treatment of cancer, they nevertheless only show a certain positive effect, particularly on middle- and late-stage cancer [88], which is due to limited specificity to malignant tumor cells/tissues and undesired adverse effects to normal tissues/organs [89]. The rise of nanotechnology offers completely new viewpoints for cancer treatment. In particular, to improve therapeutic efficacy and reduce toxic side effect, a variety of nanomaterials (e.g., metal nanoparticles, carbon-based nanomaterials, and silicon-based nanomaterials, etc.) have been explored as novel high-efficacy drug delivery systems for therapeutic agents [89–98]. In addition to chemotherapy, novel nanotechnology-based phototherapy (e.g., photodynamic therapy (PDT) [99] and photothermal therapy (PTT) [100]) have been developed for cancer treatment with encouraging therapeutic outcomes.

Silicon nanotechnology has emerged as one kind of promising means for cancer therapy, rivaling, or complementing other well-studied nanotechnologies (Fig. 1.5). For example, porous silicon (pSi) -based drug carriers have demonstrated to be highly efficacious for fighting against cancers [91]. Specifically, different types of drug molecules (e.g., doxorubicin (DOX) [101, 102], paclitaxel [103], mitoxantrone dihydrochloride (MTX) [104], camptothecin (CPT) [105], and indomethacin (IMC) [106], etc.), therapeutic genes and proteins [107–111] can be readily loaded on pSi, and then specifically target the tumor cells/tissues, followed by a continuous release of therapeutic agents, eventually leading to persistent and effective destruction of cancer cells for a long term, with minimal toxic side effect. The pSi functionalized with photosensitizers [99], radiosensitizers [112], or heat producers [113] are also explored as novel agents for high-efficacy phototherapy or radiotherapy of cancer. It is worth noting that, taking advantage of porous structures, huge surface-to-volume ratio, and large surface area, SiNWs are recently found to be novel drug nanocarriers with ultrahigh drug-loading capacity of $\sim 20,800$ mg/g [114], which is several folds larger than those ($\sim 1,200–4,000$ mg/g) reported for most nanomaterial (e.g., mesoporous silica, single-walled carbon nanotubes, graphene, etc.)-based carriers [89, 96, 115]. As a result, the resultant SiNWs-based nanocarriers are highly efficacious for in vitro and in vivo cancer therapy. Moreover, silicon-based hyperthermia nanoagents made of gold nanoparticles/nanospheres-coated SiNWs have recently been explored for cancer treatment. Specifically, the SiNWs-based hyperthermia nanoagents conjugated with antibodies are capable of simultaneously capturing and destroying malignant tumor cells. We summarize the above-mentioned progresses on development of silicon nanotechnology for cancer therapy in Chap. 5.

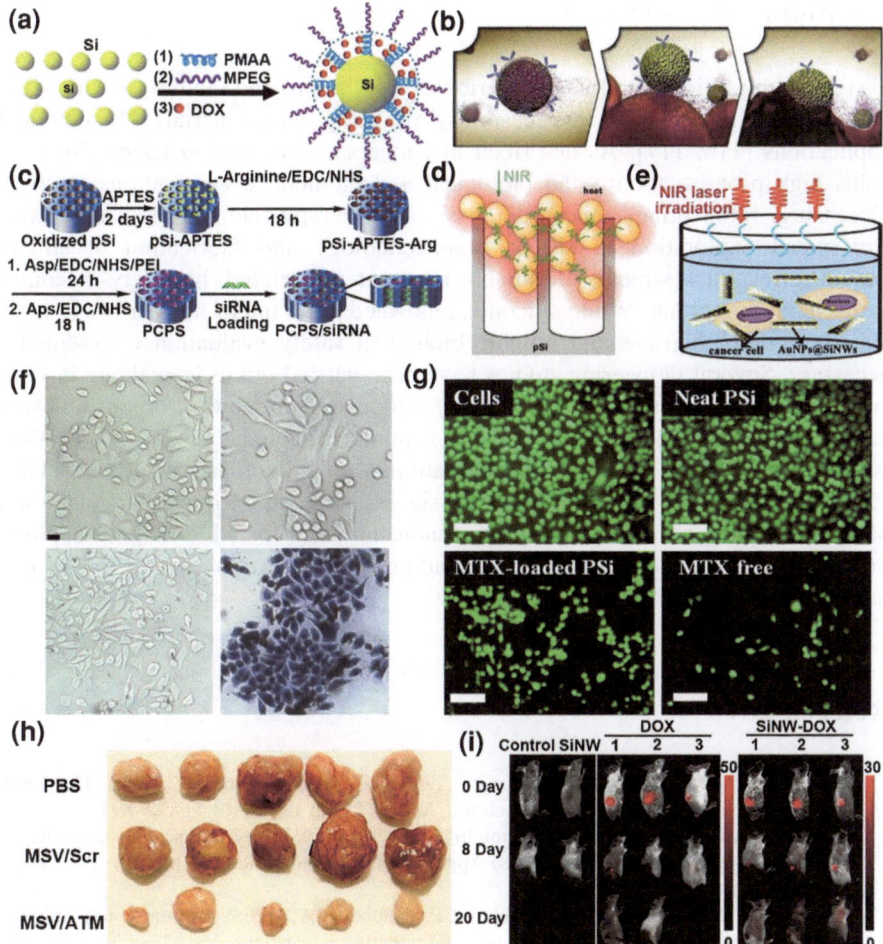

Fig. 1.5 Silicon nanobiotechnology for cancer therapy. **a** Schematic showing design of silicon-based drug carriers made of copolymer-encapsulated nanocrystalline silicon (ncSi). Reprinted with the permission from Ref. [102]. Copyright 2012 American Chemical Society. **b** Schematic illustration of pSiNP-assisted vectorization of alkaloid camptothecin to tumor cells. Reproduced from Ref. [105] by permission of John Wiley & Sons Inc. **c** Polycation-functionalized nanoporous silicon (PCPS) is employed as novel silicon-based delivery systems for gene silencing agents. Reprinted with the permission from Ref. [110]. Copyright 2013 American Chemical Society. **d** Gold nanospheres-coated SiNWs facilitate improvement of thermal efficiency. Reproduced from Ref. [113] by permission of John Wiley & Sons Inc. **e** and **f** Malignant tumor cells are efficiently killed by the silicon-based hyperthermia agents made of AuNPs-decorated SiNWs. Reprinted with the permission from Ref. [100]. Copyright 2012 American Chemical Society. **g** Fluorescence images of MDA-MB-231 cells treated by pure pSi, free MTX and pSi loaded with MTX. Green fluorescence is ascribed to fluorescein diacetate staining. Reprinted from Ref. [104], Copyright 2013, with permission from Elsevier. **h** Tumor growth in MDA-MB-231 xenograft model is significantly inhibited by the MSV/ATM (porous silicon-based multistage vector/ataxia-telangiectasia mutated). In contrast, for PBS- or MSV/Scr (scramble control siRNA)-treated mouse, the tumor size is up to 150–200 mm³. Reproduced from Ref. [111] by permission of John Wiley & Sons Inc. **i** Fluorescence images of KB bearing-nude Balb/c mice treated with PBS, free SiNWs, free DOX, and DOX-loaded SiNWs, respectively. Reproduced from Ref. [114] by permission of John Wiley & Sons Inc

1.5 Biosafety Assessment

Systematic investigation of nanomaterial-related in vitro and in vivo toxicity is the key premise to evaluate the feasibility of using nanomaterials for practical applications [116, 117]. As described in above sections, silicon nanotechnology holds high promise for myriad biological and biomedical applications, such as biosensing, bioimaging, cancer diagnosis and therapy, etc. Accompanied with widespread exploration of silicon nanotechnology and fabrication of silicon nanomaterials, assessment of silicon nanomaterial-related biosafety becomes increasingly important. While silicon is renowned as a kind of non- or lowly toxic material, comprehensive and reliable biological safety evaluation is essentially necessary. Several pioneering studies have been carried out to investigate in vitro and in vivo behaviors (e.g., cell viability, biodegradable ability, organ biodistribution, pharmacokinetics, etc.) of silicon nanomaterials (e.g., SiNPs and SiNWs), primarily suggesting excellent biocompatibility of silicon nanomaterials [118–122]. However, there still exist controversial issues in an adequate demonstration of the safety and reliability of silicon nanotechnology for wide-ranging applications, which are urgently required to be addressed in the future. A review of this field is presented in Chap. 6.

References

1. Riehemann K, Schneider SW, Luger TA, Godin B, Ferrari M, Fuchs H (2009) Nanomedicine-challenge and perspectives. Angew Chem Int Ed 48(5):872–897
2. Poole C, Owens FJ (2003) Introduction to nanotechnology. Wiley-Interscience, New Jersey
3. De M, Ghosh PS, Rotello VM (2008) Applications of nanoparticles in biology. Adv Mater 20(22):4225–4241
4. Rothenfluh DA, Bermudez H, O'Neil CP, Hubbell JA (2008) Biofunctional polymer nanoparticles for intra-articular targeting and retention in cartilage. Nat Mater 7(3):248–254
5. Hong H, Zhang Y, Sun J, Cai W (2009) Molecular imaging and therapy of cancer with radiolabeled nanoparticles. Nano Today 4(5):399–413
6. Kostarelos K, Bianco A, Prato M (2009) Promises, facts and challenges for carbon nanotubes in imaging and therapeutics. Nat Nanotechnol 4(10):627–633
7. Grom GF, Lockwood DJ, McCaffrey JP, Labbe HJ, Fauchet PM, White B, Diener J, Kovalev D, Koch F, Tsybeskov L (2000) Ordering and self-organization in nanocrystalline silicon. Nature 407(6802):358–361
8. Pavesi L, Dal Negro L, Mazzoleni C, Franzo G, Priolo F (2000) Optical gain in silicon nanocrystals. Nature 408(6811):440–444
9. Ding Z, Quinn BM, Haram SK, Pell LE, Korgel BA, Bard AJ (2002) Electrochemistry and electrogenerated chemiluminescence from silicon nanocrystal quantum dots. Science 296(5571):1293–1297
10. Schmidt V, Wittemann JV, Senz S, Gösele U (2009) Silicon nanowires: a review on aspects of their growth and their electrical properties. Adv Mater 21(25–26):2681–2702
11. He Y, Fan CH, Lee ST (2010) Silicon nanostructures for bioapplications. Nano Today 5(4):282–295

12. Peng K-Q, Wang X, Li L, Hu Y, Lee S-T (2013) Silicon nanowires for advanced energy conversion and storage. Nano Today 8(1):75–97
13. Murphy CJ, Thompson LB, Chernak DJ, Yang JA, Sivapalan ST, Boulos SP, Huang J, Alkilany AM, Sisco PN (2011) Gold nanorod crystal growth: from seed-mediated synthesis to nanoscale sculpting. Curr Opin Colloid Interface Sci 16(2):128–134
14. Dreaden EC, Alkilany AM, Huang X, Murphy CJ, El-Sayed MA (2012) The golden age: gold nanoparticles for biomedicine. Chem Soc Rev 41(7):2740–2779
15. Chen J, McLellan JM, Siekkinen A, Xiong Y, Li Z-Y, Xia Y (2006) Facile synthesis of gold-silver nanocages with controllable pores on the surface. J Am Chem Soc 128(46):14776–14777
16. Bruchez M Jr, Moronne M, Gin P, Weiss S, Alivisatos AP (1998) Semiconductor nanocrystals as fluorescent biological labels. Science 281(5385):2013–2016
17. Chan WC, Nie S (1998) Quantum dot bioconjugates for ultrasensitive nonisotopic detection. Science 281(5385):2016–2018
18. Medintz IL, Uyeda HT, Goldman ER, Mattoussi H (2005) Quantum dot bioconjugates for imaging, labelling and sensing. Nat Mater 4(6):435–446
19. Michalet X, Pinaud F, Bentolila L, Tsay J, Doose S, Li J, Sundaresan G, Wu A, Gambhir S, Weiss S (2005) Quantum dots for live cells, in vivo imaging, and diagnostics. Science 307(5709):538–544
20. Chung C, Kim Y-K, Shin D, Ryoo S-R, Hong BH, Min D-H (2013) Biomedical applications of graphene and graphene oxide. Acc Chem Res 46(10):2211–2224
21. Zhu Y, Murali S, Cai W, Li X, Suk JW, Potts JR, Ruoff RS (2010) Graphene and graphene oxide: synthesis, properties, and applications. Adv Mater 22(35):3906–3924
22. Laurent S, Forge D, Port M, Roch A, Robic C, Vander Elst L, Muller RN (2008) Magnetic iron oxide nanoparticles: synthesis, stabilization, vectorization, physicochemical characterizations, and biological applications. Chem Rev 108(6):2064–2110
23. Basak S, Chen D-R, Biswas P (2007) Electrospray of ionic precursor solutions to synthesize iron oxide nanoparticles: modified scaling law. Chem Eng Sci 62(4):1263–1268
24. Martínez-Mera I, Espinosa-Pesqueira ME, Pérez-Hernández R, Arenas-Alatorre J (2007) Synthesis of magnetite (Fe_3O_4) nanoparticles without surfactants at room temperature. Mater Lett 61(23–24):4447–4451
25. Ma D, Lee C, Au F, Tong S, Lee S (2003) Small-diameter silicon nanowire surfaces. Science 299(5614):1874–1877
26. Wilson WL, Szajowski P, Brus L (1993) Quantum confinement in size-selected, surface-oxidized silicon nanocrystals. Science 262:1242
27. Park N-M, Choi C-J, Seong T-Y, Park S-J (2001) Quantum confinement in amorphous silicon quantum dots embedded in silicon nitride. Phys Rev Lett 86(7):1355–1357
28. Yang C-S, Bley RA, Kauzlarich SM, Lee HW, Delgado GR (1999) Synthesis of alkyl-terminated silicon nanoclusters by a solution route. J Am Chem Soc 121(22):5191–5195
29. Jurbergs D, Rogojina E, Mangolini L, Kortshagen U (2006) Silicon nanocrystals with ensemble quantum yields exceeding 60 %. Appl Phys Lett 88(23):233116
30. Kang Z, Liu Y, Tsang CHA, Ma DDD, Fan X, Wong NB, Lee ST (2009) Water-soluble silicon quantum dots with wavelength-tunable photoluminescence. Adv Mater 21(6):661–664
31. He Y, Zhong Y, Peng F, Wei X, Su Y, Lu Y, Su S, Gu W, Liao L, Lee S-T (2011) One-pot microwave synthesis of water-dispersible, ultraphoto-and pH-stable, and highly fluorescent silicon quantum dots. J Am Chem Soc 133(36):14192–14195
32. Zhong Y, Peng F, Wei X, Zhou Y, Wang J, Jiang X, Su Y, Su S, Lee ST, He Y (2012) Microwave-assisted synthesis of biofunctional and fluorescent silicon nanoparticles using proteins as hydrophilic ligands. Angew Chem Int Ed 51(34):8485–8489
33. Zhong Y, Peng F, Bao F, Wang S, Ji X, Yang L, Su Y, Lee S-T, He Y (2013) Large-scale aqueous synthesis of fluorescent and biocompatible silicon nanoparticles and their use as highly photostable biological probes. J Am Chem Soc 135(22):8350–8356

34. Morales AM, Lieber CM (1998) A laser ablation method for the synthesis of crystalline semiconductor nanowires. Science 279(5348):208–211
35. Hochbaum AI, Fan R, He R, Yang P (2005) Controlled growth of Si nanowire arrays for device integration. Nano Lett 5(3):457–460
36. Chung S-W, Yu J-Y, Heath JR (2000) Silicon nanowire devices. Appl Phys Lett 76(15):2068–2070
37. Lew K-K, Redwing JM (2003) Growth characteristics of silicon nanowires synthesized by vapor-liquid-solid growth in nanoporous alumina templates. J Cryst Growth 254(1–2):14–22
38. Wu Y, Cui Y, Huynh L, Barrelet CJ, Bell DC, Lieber CM (2004) Controlled growth and structures of molecular-scale silicon nanowires. Nano Lett 4(3):433–436
39. Zhang RQ, Lifshitz Y, Lee ST (2003) Oxide-assisted growth of semiconducting nanowires. Adv Mater 15(7–8):635–640
40. Peng KQ, Yan YJ, Gao SP, Zhu J (2002) Synthesis of large-area silicon nanowire arrays via self-assembling nanoelectrochemistry. Adv Mater 14(16):1164–1167
41. Peng K, Xu Y, Wu Y, Yan Y, Lee S-T, Zhu J (2005) Aligned single-crystalline Si nanowire arrays for photovoltaic applications. Small 1(11):1062–1067
42. Peng K, Lu A, Zhang R, Lee S-T (2008) Motility of metal nanoparticles in silicon and induced anisotropic silicon etching. Adv Funct Mater 18(19):3026–3035
43. He Y, Zhong Y, Peng F, Wei X, Su Y, Su S, Gu W, Liao L, Lee ST (2011) Highly luminescent water-dispersible silicon nanowires for long-term immunofluorescent cellular imaging. Angew Chem Int Ed 50:3080–3083
44. Diamond D (1998) In: Diamond D (ed) Principles of chemical and biological sensors. John Wiley & Sons, New York, pp 1–18
45. Daniel M-C, Astruc D (2004) Gold nanoparticles: assembly, supramolecular chemistry, quantum-size-related properties, and applications toward biology, catalysis, and nanotechnology. Chem Rev 104(1):293–346
46. Rosi NL, Mirkin CA (2005) Nanostructures in biodiagnostics. Chem Rev 105(4):1547–1562
47. Patolsky F, Lieber CM (2005) Nanowire nanosensors. Mater Today 8(4):20–28
48. Patolsky F, Timko BP, Yu G, Fang Y, Greytak AB, Zheng G, Lieber CM (2006) Detection, stimulation, and inhibition of neuronal signals with high-density nanowire transistor arrays. Science 313(5790):1100–1104
49. Patolsky F, Zheng G, Lieber CM (2006) Fabrication of silicon nanowire devices for ultrasensitive, label-free, real-time detection of biological and chemical species. Nat Protocols 1(4):1711–1724
50. J-i Hahm, Lieber CM (2004) Direct ultrasensitive electrical detection of DNA and DNA sequence variations using nanowire nanosensors. Nano Lett 4(1):51–54
51. Gao Z, Agarwal A, Trigg AD, Singh N, Fang C, Tung C-H, Fan Y, Buddharaju KD, Kong J (2007) Silicon nanowire arrays for label-free detection of DNA. Anal Chem 79(9):3291–3297
52. Cattani-Scholz A, Pedone D, Dubey M, Neppl S, Nickel B, Feulner P, Schwartz J, Abstreiter G, Tornow M (2008) Organophosphonate-based PNA-functionalization of silicon nanowires for label-free DNA detection. ACS Nano 2(8):1653–1660
53. Gao A, Zou N, Dai P, Lu N, Li T, Wang Y, Zhao J, Mao H (2013) Signal-to-noise ratio enhancement of silicon nanowires biosensor with rolling circle amplification. Nano Lett 13(9):4123–4130
54. Cui Y, Wei Q, Park H, Lieber CM (2001) Nanowire nanosensors for highly sensitive and selective detection of biological and chemical species. Science 293(5533):1289–1292
55. Zheng G, Patolsky F, Cui Y, Wang WU, Lieber CM (2005) Multiplexed electrical detection of cancer markers with nanowire sensor arrays. Nat Biotechnol 23(10):1294–1301
56. Patolsky F, Zheng G, Hayden O, Lakadamyali M, Zhuang X, Lieber CM (2004) Electrical detection of single viruses. Proc Natl Acad Sci USA 101(39):14017–14022

57. Lin S-P, Pan C-Y, Tseng K-C, Lin M-C, Chen C-D, Tsai C-C, Yu S-H, Sun Y-C, Lin T-W, Chen Y-T (2009) A reversible surface functionalized nanowire transistor to study protein–protein interactions. Nano Today 4(3):235–243

58. Su S, He Y, Song S, Li D, Wang L, Fan C, Lee S-T (2010) A silicon nanowire-based electrochemical glucose biosensor with high electrocatalytic activity and sensitivity. Nanoscale 2(9):1704–1707

59. Shao MW, Shan YY, Wong NB, Lee ST (2005) Silicon nanowire sensors for bioanalytical applications: glucose and hydrogen peroxide detection. Adv Funct Mater 15(9):1478–1482

60. Leng W, Yasseri AA, Sharma S, Li Z, Woo HY, Vak D, Bazan GC, Kelley AM (2006) Silver nanocrystal-modified silicon nanowires as substrates for surface-enhanced Raman and hyper-Raman scattering. Anal Chem 78(17):6279–6282

61. Zhang B, Wang H, Lu L, Ai K, Zhang G, Cheng X (2008) Large-area silver-coated silicon nanowire arrays for molecular sensing using surface-enhanced Raman spectroscopy. Adv Funct Mater 18(16):2348–2355

62. Wang H, Han X, Ou X, Lee C-S, Zhang X, Lee S-T (2013) Silicon nanowire based single-molecule SERS sensor. Nanoscale 5(17):8172–8176

63. Wei X, Su S, Guo Y, Jiang X, Zhong Y, Su Y, Fan C, Lee S-T, He Y (2013) A molecular beacon-based signal-off surface-enhanced Raman scattering strategy for highly sensitive, reproducible, and multiplexed DNA detection. Small 9(15):2493–2499

64. Vendrell M, Maiti KK, Dhaliwal K, Chang Y-T (2013) Surface-enhanced Raman scattering in cancer detection and imaging. Trends Biotechnol 31(4):249–257

65. Wang Y, Yan B, Chen L (2013) SERS tags: novel optical nanoprobes for bioanalysis. Chem Rev 113(3):1391–1428

66. He Y, Su S, Xu T, Zhong Y, Zapien JA, Li J, Fan C, Lee S-T (2011) Silicon nanowires-based highly-efficient SERS-active platform for ultrasensitive DNA detection. Nano Today 6(2):122–130

67. Han X, Wang H, Ou X, Zhang X (2012) Highly sensitive, reproducible, and stable SERS sensors based on well-controlled silver nanoparticle-decorated silicon nanowire building blocks. J Mater Chem 22(28):14127–14132

68. Jiang Z, Jiang X, Su S, Wei X, Lee S, He Y (2012) Silicon-based reproducible and active surface-enhanced Raman scattering substrates for sensitive, specific, and multiplex DNA detection. Appl Phys Lett 100(20):203104

69. Jiang X, Jiang Z, Xu T, Su S, Zhong Y, Peng F, Su Y, He Y (2013) Surface-enhanced Raman scattering-based sensing in vitro: facile and label-free detection of apoptotic cells at the single-cell level. Anal Chem 85(5):2809–2816

70. Su S, Wei XP, Zhong YL, Guo YY, Su YY, Huang Q, Lee ST, Fan CH, He Y (2012) Silicon nanowire-based molecular beacons for high-sensitivity and sequence-specific DNA multiplexed analysis. ACS Nano 6(3):2582–2590

71. Gao X, Cui Y, Levenson RM, Chung LWK, Nie S (2004) In vivo cancer targeting and imaging with semiconductor quantum dots. Nat Biotechnol 22(8):969–976

72. Sun C, Lee JS, Zhang M (2008) Magnetic nanoparticles in MR imaging and drug delivery. Adv Drug Del Rev 60(11):1252–1265

73. Gao J, Gu H, Xu B (2009) Multifunctional magnetic nanoparticles: design, synthesis, and biomedical applications. Acc Chem Res 42(8):1097–1107

74. Brähler M, Georgieva R, Buske N, Müller A, Müller S, Pinkernelle J, Teichgräber U, Voigt A, Bäumler H (2006) Magnetite-loaded carrier erythrocytes as contrast agents for magnetic resonance imaging. Nano Lett 6(11):2505–2509

75. Su YY, He Y, Lu HT, Sai LM, Li QN, Li WX, Wang LH, Shen PP, Huang Q, Fan CH (2009) The cytotoxicity of cadmium based, aqueous phase-synthesized, quantum dots and its modulation by surface coating. Biomaterials 30(1):19–25

76. Choi HS, Liu W, Misra P, Tanaka E, Zimmer JP, Ipe BI, Bawendi MG, Frangioni JV (2007) Renal clearance of quantum dots. Nat Biotechnol 25(10):1165–1170

77. Park J-H, Gu L, von Maltzahn G, Ruoslahti E, Bhatia SN, Sailor MJ (2009) Biodegradable luminescent porous silicon nanoparticles for in vivo applications. Nat Mater 8(4):331–336

78. Li Z, Ruckenstein E (2004) Water-soluble poly (acrylic acid) grafted luminescent silicon nanoparticles and their use as fluorescent biological staining labels. Nano Lett 4(8):1463–1467

79. He Y, Kang ZH, Li QS, Tsang CHA, Fan CH, Lee ST (2009) Ultrastable, highly fluorescent, and water-dispersed silicon-based nanospheres as cellular probes. Angew Chem Int Ed 48:128–132

80. He Y, Su Y, Yang X, Kang Z, Xu T, Zhang R, Fan C, Lee S-T (2009) Photo and pH stable, highly-luminescent silicon nanospheres and their bioconjugates for immunofluorescent cell imaging. J Am Chem Soc 131(12):4434–4438

81. Erogbogbo F, Yong K-T, Roy I, Hu R, Law W-C, Zhao W, Ding H, Wu F, Kumar R, Swihart MT, Prasad PN (2011) In vivo targeted cancer imaging, sentinel lymph node mapping and multi-channel imaging with biocompatible silicon nanocrystals. ACS Nano 5(1):413–423

82. Ohta S, Shen P, Inasawa S, Yamaguchi Y (2012) Size-and surface chemistry-dependent intracellular localization of luminescent silicon quantum dot aggregates. J Mater Chem 22(21):10631–10638

83. Choi HS, Liu W, Liu F, Nasr K, Misra P, Bawendi MG, Frangioni JV (2009) Design considerations for tumour-targeted nanoparticles. Nat Nanotechnol 5(1):42–47

84. Ackerman JJ (2013) Magnetic resonance imaging: silicon for the future. Nat Nanotechnol 8(5):313–315

85. Atkins TM, Cassidy MC, Lee M, Ganguly S, Marcus CM, Kauzlarich SM (2013) Synthesis of long T1 silicon nanoparticles for hyperpolarized ^{29}Si magnetic resonance imaging. ACS Nano 7(2):1609–1617

86. Cassidy MC, Chan HR, Ross BD, Bhattacharya PK, Marcus CM (2013) In vivo magnetic resonance imaging of hyperpolarized silicon particles. Nat Nanotechnol 8(5):363–368

87. Tu C, Ma X, Pantazis P, Kauzlarich SM, Louie AY (2010) Paramagnetic, silicon quantum dots for magnetic resonance and two-photon imaging of macrophages. J Am Chem Soc 132(6):2016–2023

88. Iyer AK, Singh A, Ganta S, Amiji MM (2013) Role of integrated cancer nanomedicine in overcoming drug resistance. Adv Drug Deliver Rev 65(13–14):1784–1802

89. Yang K, Feng L, Shi X, Liu Z (2013) Nano-graphene in biomedicine: theranostic applications. Chem Soc Rev 42(2):530–547

90. Xia Y, Li W, Cobley CM, Chen J, Xia X, Zhang Q, Yang M, Cho EC, Brown PK (2011) Gold nanocages: from synthesis to theranostic applications. Acc Chem Res 44(10):914–924

91. Anglin EJ, Cheng L, Freeman WR, Sailor MJ (2008) Porous silicon in drug delivery devices and materials. Adv Drug Deliver Rev 60(11):1266–1277

92. Gultepe E, Nagesha D, Sridhar S, Amiji M (2010) Nanoporous inorganic membranes or coatings for sustained drug delivery in implantable devices. Adv Drug Deliver Rev 62(3):305–315

93. Jaganathan H, Godin B (2012) Biocompatibility assessment of Si-based nano- and micro-particles. Adv Drug Deliver Rev 64(15):1800–1819

94. Slowing II, Vivero-Escoto JL, Wu C-W, Lin VSY (2008) Mesoporous silica nanoparticles as controlled release drug delivery and gene transfection carriers. Adv Drug Deliver Rev 60(11):1278–1288

95. Chen Y, Chen H, Shi J (2013) In vivo vivo-safety evaluations and diagnostic/therapeutic applications of chemically designed mesoporous silica nanoparticles. Adv Mater 25(23):3144–3176

96. Tang F, Li L, Chen D (2012) Mesoporous silica nanoparticles: synthesis, biocompatibility and drug delivery. Adv Mater 24(12):1504–1534

97. Elsabahy M, Wooley KL (2012) Design of polymeric nanoparticles for biomedical delivery applications. Chem Soc Rev 41(7):2545–2561

98. Hubbell JA, Chilkoti A (2012) Nanomaterials for drug delivery. Science 337(6092):303–305

99. Xiao L, Gu L, Howell SB, Sailor MJ (2011) Porous silicon nanoparticle photosensitizers for singlet oxygen and their phototoxicity against cancer cells. ACS Nano 5(5):3651–3659

100. Su Y, Wei X, Peng F, Zhong Y, Lu Y, Su S, Xu T, Lee S-T, He Y (2012) Gold nanoparticles-decorated silicon nanowires as highly efficient near-infrared hyperthermia agents for cancer cells destruction. Nano Lett 12(4):1845–1850

101. Gu L, Park J-H, Duong KH, Ruoslahti E, Sailor MJ (2010) Magnetic luminescent porous silicon microparticles for localized delivery of molecular frug payloads. Small 6(22):2546–2552

102. Xu Z, Wang D, Guan M, Liu X, Yang Y, Wei D, Zhao C, Zhang H (2012) Photoluminescent silicon nanocrystal-based multifunctional carrier for pH-regulated drug delivery. ACS Appl Mater Interfaces 4(7):3424–3431

103. Shen H, Rodriguez-Aguayo C, Xu R, Gonzalez-Villasana V, Mai J, Huang Y, Zhang G, Guo X, Bai L, Qin G, Deng X, Li Q, Erm DR, Aslan B, Liu X, Sakamoto J, Chavez-Reyes A, Han H-D, Sood AK, Ferrari M, Lopez-Berestein G (2013) Enhancing chemotherapy response with sustained EphA2 silencing using multistage vector delivery. Clin Cancer Res 19(7):1806–1815

104. Tzur-Balter A, Gilert A, Massad-Ivanir N, Segal E (2013) Engineering porous silicon nanostructures as tunable carriers for mitoxantrone dihydrochloride. Acta Biomater 9(4):6208–6217

105. Secret E, Smith K, Dubljevic V, Moore E, Macardle P, Delalat B, Rogers M-L, Johns TG, Durand J-O, Cunin F, Voelcker NH (2013) Antibody-functionalized porous silicon nanoparticles for vectorization of hydrophobic drugs. Adv Health Mater 2(5):718–727

106. Liu D, Bimbo LM, Mäkilä E, Villanova F, Kaasalainen M, Herranz-Blanco B, Caramella CM, Lehto V-P, Salonen J, Herzig K-H, Hirvonen J, Santos HA (2013) Co-delivery of a hydrophobic small molecule and a hydrophilic peptide by porous silicon nanoparticles. J Control Release 170(2):268–278

107. Kinnari PJ, Hyvönen MLK, Mäkilä EM, Kaasalainen MH, Rivinoja A, Salonen JJ, Hirvonen JT, Laakkonen PM, Santos HA (2013) Tumour homing peptide-functionalized porous silicon nanovectors for cancer therapy. Biomaterials 34(36):9134–9141

108. Tanaka T, Mangala LS, Vivas-Mejia PE, Nieves-Alicea R, Mann AP, Mora E, Han H-D, Shahzad MM, Liu X, Bhavane R, Gu J, Fakhoury JR, Chiappini C, Lu C, Matsuo K, Godin B, Stone RL, Nick AM, Lopez-Berestein G, Sood AK, Ferrari M (2010) Sustained small interfering RNA delivery by mesoporous silicon particles. Cancer Res 70(9):3687–3696

109. Ferrari M (2010) Experimental therapies: vectoring siRNA therapeutics into the clinic. Nat Rev Clin Oncol 7(9):485–486

110. Shen J, Xu R, Mai J, Kim H-C, Guo X, Qin G, Yang Y, Wolfram J, Mu C, Xia X, Gu J, Liu X, Mao Z-W, Ferrari M, Shen H (2013) High capacity nanoporous silicon carrier for systemic delivery of gene silencing therapeutics. ACS Nano 7(11):9867–9880

111. Xu R, Huang Y, Mai J, Zhang G, Guo X, Xia X, Koay EJ, Qin G, Erm DR, Li Q, Liu X, Ferrari M, Shen H (2013) Multistage vectored siRNA targeting ataxia-telangiectasia mutated for breast cancer therapy. Small 9(9–10):1799–1808

112. David Gara P, Garabano N, Llansola Portoles M, Moreno MS, Dodat D, Casas O, Gonzalez M, Kotler M (2012) ROS enhancement by silicon nanoparticles in X-ray irradiated aqueous suspensions and in glioma C6 cells. J Nanopart Res 14(3):1–13

113. Shen H, You J, Zhang G, Ziemys A, Li Q, Bai L, Deng X, Erm DR, Liu X, Li C, Ferrari M (2012) Cooperative, nanoparticle-enabled thermal therapy of breast cancer. Adv Health Mater 1(1):84–89

114. Peng F, Su Y, Wei X, Lu Y, Zhou Y, Zhong Y, Lee S-T, He Y (2013) Silicon-nanowire-based nanocarriers with ultrahigh drug-loading capacity for in vitro and in vivo cancer therapy. Angew Chem Int Ed 52(5):1457–1461

115. Liu Z, Sun X, Nakayama-Ratchford N, Dai H (2007) Supramolecular chemistry on water-soluble carbon nanotubes for drug loading and delivery. ACS Nano 1(1):50–56

116. Lewinski N, Colvin V, Drezek R (2008) Cytotoxicity of nanoparticles. Small 4(1):26–49

117. Sharifi S, Behzadi S, Laurent S, Forrest ML, Stroeve P, Mahmoudi M (2012) Toxicity of nanomaterials. Chem Soc Rev 41(6):2323–2343
118. Bimbo LM, Sarparanta M, Santos HA, Airaksinen AJ, Mäkilä E, Laaksonen T, Peltonen L, Lehto V-P, Hirvonen J, Salonen J (2010) Biocompatibility of thermally hydrocarbonized porous silicon nanoparticles and their biodistribution in rats. ACS Nano 4(6):3023–3032
119. Ohta S, Inasawa S, Yamaguchi Y (2012) Real time observation and kinetic modeling of the cellular uptake and removal of silicon quantum dots. Biomaterials 33(18):4639–4645
120. Ohta S, Shen P, Inasawa S, Yamaguchi Y (2012) Size- and surface chemistry-dependent intracellular localization of luminescent silicon quantum dot aggregates. J Mater Chem 22(21):10631–10638
121. Qi S, Yi C, Ji S, Fong C-C, Yang M (2009) Cell adhesion and spreading behavior on vertically aligned silicon nanowire arrays. ACS Appl Mater Interfaces 1(1):30–34
122. Roberts JR, Mercer RR, Chapman RS, Cohen GM, Bangsaruntip S, Schwegler-Berry D, Scabilloni JF, Castranova V, Antonini JM, Leonard SS (2012) Pulmonary toxicity, distribution, and clearance of intratracheally instilled silicon nanowires in rats. J Nanomater 2012:17

Chapter 2
Silicon Nanostructures

Abstract Functional nanomaterials play fundamental roles in the development of nanotechnology, serving as novel and powerful tools for both basic studies and practical applications. Silicon nanomaterials are an important type of nanomaterials, exhibiting unique optical, electronic, or/and mechanical properties. The fast development of silicon nanomaterials with well-defined structures and required functionalities has vastly promoted the advancement of silicon nanotechnology. Silicon nanoparticles (SiNPs) and silicon nanowires (SiNWs) are well known as the most important zero- and one-dimensional silicon nanostructures. In the past three decades, scientists have made great strides in developing a great deal of fabrication techniques to prepare SiNPs and SiNWs. In particular, solution-phase reduction, electrochemical etching and microwave-assisted synthesis, etc., have been well developed for the production of SiNPs. On the other hand, several well-studied strategies (e.g., chemical vapor deposition (CVD), oxide-assisted growth (OAG), electroless etching, etc.) are highly efficacious for the synthesis of SiNWs. In this chapter, we give an introduction to these classic synthetic methods in a detailed way, and discuss the prospect of the design and fabrication of functional silicon nanostructures.

Keywords Silicon nanostructures · Nanotechnology · Synthetic methods · Surface modification · Silicon nanoparticles · Silicon nanowires · Silicon nanohybrids

Increasingly rapid advancement of nanotechnology has fueled a continual and urgent investigation for fabricating functional nanomaterials [1, 2]. Scientists have done elegant work on rational design of nanomaterials with well-defined structures [3–6]. To date, a variety of nanomaterials (e.g., silver/gold nanoparticles, magnetic nanoparticles, fluorescent semiconductor quantum dots, carbon nanodots/nanotubes, graphene, etc.) have been well established, greatly promoting the development of chemical, physical, and biological fields from basic research to practical applications.

Silicon material is the leading semiconductor material and dominates current industry. Silicon nanomaterials, known as one of the most important types of nanomaterials, feature a number of unique merits, such as excellent

Y. He and Y. Su, *Silicon Nano-biotechnology*, SpringerBriefs in Molecular Science, DOI: 10.1007/978-3-642-54668-6_2, © The Author(s) 2014

electronic/mechanical/optical properties, huge surface-to-volume ratios, and facile surface modification [7–10]. More importantly, the renowned biocompability of silicon (e.g., silicon naturally exists in human as a common trace element) leads to the promising prospect of silicon nanomaterials-based applications, including solar cells, sensors, catalysis, bioimaging, etc. To meet the increasing requirement of these applications, silicon nanomaterials with controllable structures and required functionality remain in great demand.

This chapter mainly introduces representative strategies for preparation of the most important zero- and one-dimensional silicon nanostructures, i.e., silicon nanoparticles (SiNPs) and silicon nanowires (SiNWs). Typically, several dominant synthetic methods for the preparation and surface modification of SiNPs are illustrated in Sect. 2.1. In Sect. 2.2, we introduce the classic strategies for synthesis of SiNWs, and further discuss the pros and cons of these methods. We then give a brief introduction to silicon-based nanohybrids in Sect. 2.3, and finally make a summary of this chapter in Sect. 2.4.

2.1 Fluorescent Silicon Nanoparticles

Bulk silicon with an indirect band gap basically features poor optical properties [11]. However, when the size of silicon particles is reduced to nanoscale (generally less than 5 nm), the overlap of the electron and hole wave functions is distinctly increased, leading to dramatic enhancement of recombination rates of electrons and holes. As a result, such small-sized SiNPs exhibit relatively strong fluorescence, showing the prospect of long-awaited optical applications [12, 13]. Therefore, intense studies have been intrigued to develop fluorescent SiNPs and their optics-relative applications since the first observation of porous silicon-based fluorescence [14, 15].

In recent two decades, many synthetic strategies (e.g., solution-phase reduction [16, 17], microemulsion [18, 19], sonochemical synthesis [20], mechanochemical synthesis [21], laser ablation [22], plasma-assisted aerosol precipitation [23–25], electrochemical etching [26–30], and microwave-assisted synthesis [31–34], etc.) have been developed for the preparation of SiNPs. For the solution-phase reduction synthesis proposed by Kauzlarich and coworkers, silicon halides (e.g., $SiCl_4$) are reduced in organic solution (e.g., ethylene glycol dimethyl ether) to produce silicon nanocrystalline under mild conditions [16]. Moreover, surface of the prepared SiNPs is readily modified, offering possibility to modify the SiNPs to meet various requirements [16]. Plasma-assisted aerosol precipitation is considered as another established method for production of SiNPs with high efficiency and yield [23–25]. In this synthetic strategy, luminescent SiNPs between 2 and 8 nm can be rapidly synthesized via a single-step non-thermal plasma process [24]. In 2009, Lee and coworkers developed a polyoxometalate (POMs)-assisted electrochemical etching method for synthesizing multi-color luminescent SiNPs [27]. In this system, graphite and silicon wafer serve as anode and cathode in an electrochemical

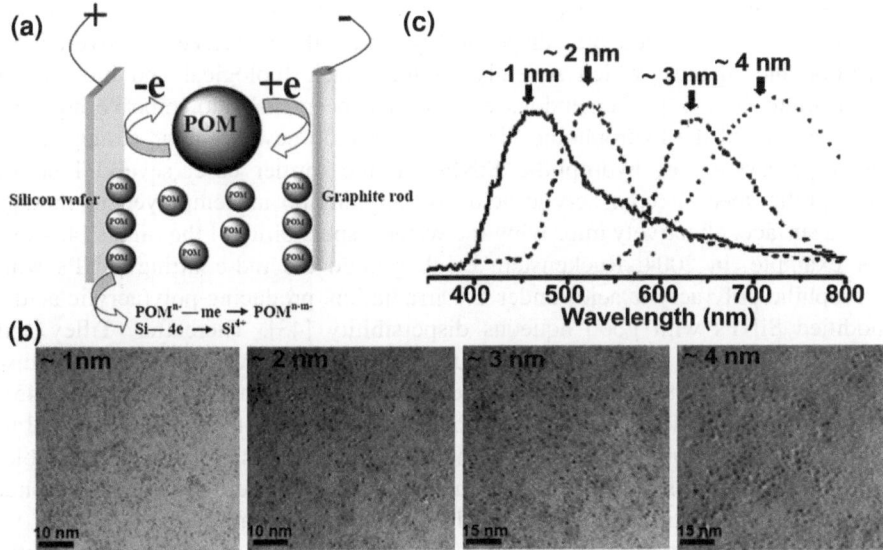

Fig. 2.1 a Schematic mode of preparation of fluorescent SiNPs through the electrochemical etching method. **b** Transmission electron microscopy (TEM) images and **c** representative PL spectra of the prepared SiNPs with controllable sizes ranging from ∼1 to ∼4 nm. Reprinted with permission from Ref. [27]. Copyright 2007 American Chemical Society

cell, respectively; besides, $H_3PMo_{12}O_{40}$ (POM) and H_2O_2 (POM + H_2O_2 = HPOM) are used as catalysts (Fig. 2.1a). Notably, the sizes of SiNPs are readily controllable via adjustment of current density, vastly facilitating the production of multi-color luminescent SiNPs with different sizes (Fig. 2.1b). Figure 2.1c displays SiNPs with sizes ranging from ∼1 to ∼4 nm, which feature wide-ranging emission spectra covering 450–740 nm. It is worthwhile to point out, in comparison to strong fluorescence of direct-band-gap semiconductors (e.g., CdTe and CdTe/CdS/ZnS QDs with high quantum yield (QY) of 60–80 %, [6, 7]), most of the prepared fluorescent SiNPs show relatively low QY (often lower than 20 %), which is possibly due to surface oxidation [35–39]. In 2006, Kortshagen et al. developed a plasma-assisted synthesis method to efficiently protect surface oxidation of SiNPs, yielding the highly luminescent SiNPs with a high QY value larger than 60 % [23]. Very recently, Li, He and coworkers reported a class of SiNPs with ultrabright photoluminescence, whose quantum yield was remarkably as high as 75 %. Specifically, the SiNPs were synthesized by the solution-reduction method, followed by modifying with diphenylamine (di) and carbazole (ca). After such a surface modification, the optical properties of SiNPs were significantly changed, e.g., the maximum emission peak of carbazole-modified SiNPs (ca-SiNPs) shifted from 405 to 480 nm, and more significantly, the photoluminescence quantum yield (PLQY) was remarkably enhanced up to 75 % [40].

It is worth pointing out that, SiNPs prepared via the above-mentioned methods often possess poor aqueous dispersibility since their surface is covered by hydrophobic ligands, which severely hampers their biological and biomedical applications [41, 42]. Tremendous efforts have been made to improve aqueous dispersibility of the hydrophobic SiNPs and develop new synthetic strategies for direct preparation of hydrophilic SiNPs. In the former case, several kind of hydrophilic species (e.g., acrylic acid and allylamine) are employed to modify SiNPs surface, effectively improving the water dispersibility of the SiNPs [43–45]. For example, in 2004, Ruckenstein et al. grafted the red-emitting SiNPs with hydrophilic poly(acrylic acid) under UV irradiation, producing poly(acrylic acid)-modified SiNPs with good aqueous dispersibility [43]. Thereafter, Tilley and coworkers reported the synthesis of blue-emitting SiNPs with good aqueous dispersibility due to surface-covered water-soluble allylamine molecules [18, 45]. Swihart's group functionalized multi-color fluorescent SiNPs with acrylic acid to render them hydrophilic [44]. While these resultant SiNPs are water-dispersible, their fluorescence is often severely quenched when pH changed, which is not suitable for applications in complicated biological environments.

To improve pH stability of the SiNPs, He et al. and Swihart et al. independently reported polymer-coated or micelle-encapsulated water-dispersible SiNPs featuring robust pH stability [29, 30, 46]. Typically, the micelle-encapsulated SiNPs maintained stable fluorescence in acidic-to-basic pH environments (pH 2–12). However, the QY value of such micelle-encapsulated SiNPs reduced to 2–4 % from ~ 17 % of pure SiNPs, since SiNPs surface was deteriorated during micelle encapsulation procedure [46]. In the case of polymer-coated SiNPs, different SiNPs were linked together by polymer chains under light irradiation, producing silicon nanospheres (SiNSs) containing tens to hundreds of SiNPs [29]. Figure 2.2a presents typical TEM images showing that the prepared nanospheres exhibit spherical structure with controllable diameters of approximately 59, 121, and 207 nm. The high-resolution TEM images (Fig. 2.2b) reveal the SiNP inside the nanosphere have the characteristic silicon nanocrystal lattice and high crystallinity. Notably, the authors demonstrated that the optical properties of the nanospheres were improved (Fig. 2.2c), whose quantum yield reached 20–25 %. On the basis of which, the same authors further introduced a kind of water-dispersed oxidized SiNSs (O-SiNSs) via thermal oxidation of the precursor SiNSs [30]. Significantly, in addition to strong fluorescence, the prepared O-SiNSs was ultrahighly stable under high-power UN irradiation (Fig. 2.2d) and in acidic-to-basic environments covering pHs 2–12 (Fig. 2.2e).

It is worth pointing out that, despite distinctly improved pH stability, the polymer-coated or micelle-encapsulated SiNPs are relatively not suitable for bioapplication due to relatively large sizes (50–200 nm) [29, 30, 46] (recent reports reveal that small-sized (<10 nm) nanoparticles are more readily to be excluded via renal clearance [47, 48]). Consequently, small-sized, fluorescent, and water-dispersible SiNPs are required for broad bioapplications. Recently, He's group presented a microwave-assisted one-pot method for synthesizing water-dispersible SiNPs using SiNWs and glutaric acid as reaction precursors [31]. In

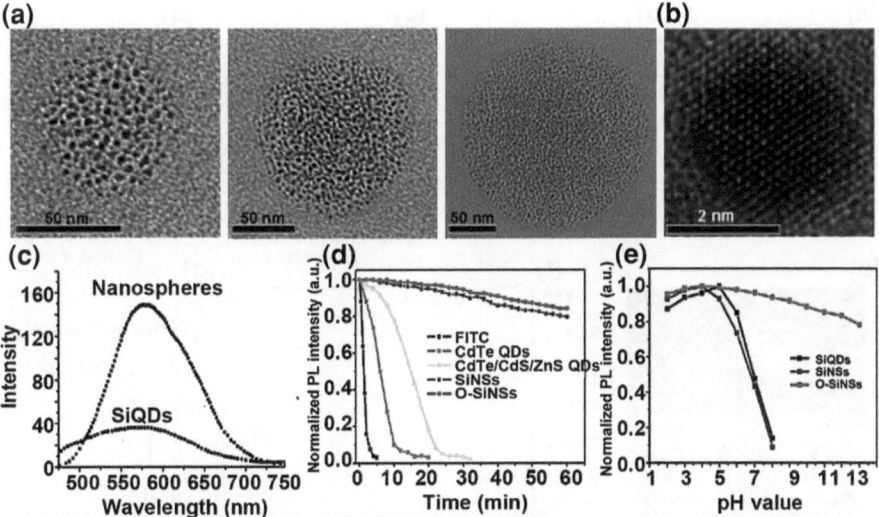

Fig. 2.2 **a** TEM images of three kinds of SiNWs with sizes of ~60 nm (*left*), 120 nm (*middle*), and 200 nm (*right*). **b** HRTEM image of a single SiNP inside the prepared SiNSs. **c** Comparison of Fluorescent intensities of the resultant SiNSs and pure SiNPs. **a–c** Reproduced from ref. [29] by permission of John Wiley & Sons Inc. **d** Temporal evolution of photoluminescence intensity of FITC, II/VI QDs, the prepared SiNSs, and O-SiNSs under long-term UV irradiation or **e** various pH values. **d–e** Reprinted with permission from Ref. [30]. Copyright 2009 American Chemical Society

this strategy, microwave dielectric heating was utilized to take advantage of its rapid temperature elevation, homogenous heating and high reaction selectivity. Notably, the prepared SiNPs featured excellent aqueous dispersibility, small sizes (~4 nm), robust pH- (pHs 1–10) and photo-stability, and strong fluorescence (~15–20 %) (Fig. 2.3). Thereafter, they further employed hydrophilic proteins (e.g., goat anti-mouse immunoglobulin) as novel ligands to prepare fluorescent SiNPs [33]. Significantly, the prepared SiNPs simultaneously possessed good aqueous dispersibility and biospecific properties owing to plenty of surface-covered protein ligands. Therefore, the as-prepared fluorescent SiNPs with bio-specific properties was efficacious for immunofluorescent cellular targeting.

For all the above "top-down" strategies, two independent procedures are often required, that is, hydrophobic SiNPs are first prepared using large-size silicon source (e.g., crystalline Si particles, SiOx powders, bulk silicon, or SiNWs, etc.), followed by surface modification with hydrophilic ligands, which nevertheless involves relatively complicated and time-consuming procedures and sometimes degrades the optical properties of SiNPs. More recently, He, Lee, and coworkers developed a facile "bottom-up" strategy for preparing fluorescent water-dispersible SiNPs by using silicon-based organic molecules (e.g., 3-(aminopropyl) trimethoxysilane, $C_6H_{17}NO_3Si$) as the reaction precursor (Fig. 2.4) [34]. The highly luminescent (PLQY: 20–25 %), water-dispersible and ultrasmall (diameter:

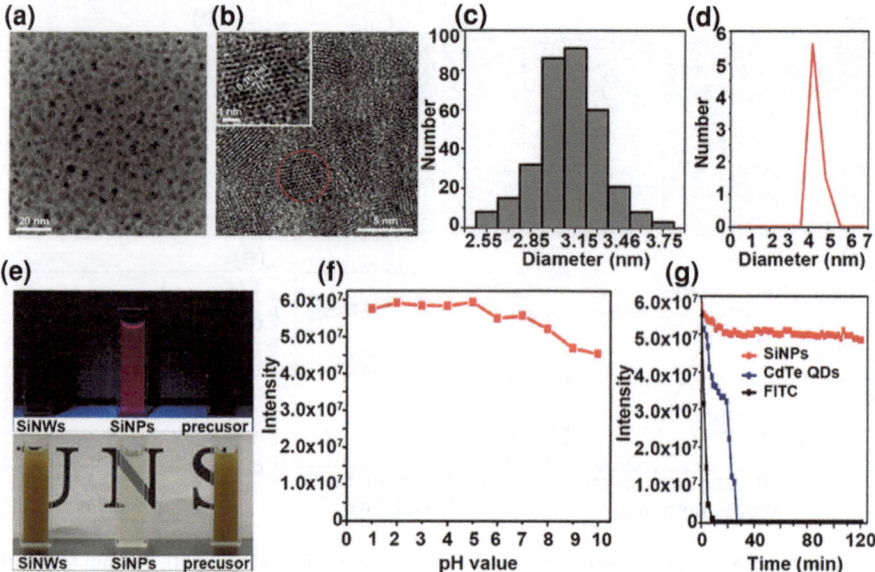

Fig. 2.3 **a** TEM and **b** HRTEM images, **c** size distribution, and **d** representative dynamic light scattering (DLS) histogram of the as-prepared SiNPs. Inset in (**b**) presents the HRTEM image of a single SiNP. **e** Photos of the SiNWs (*left*), fluorescent SiNPs (*middle*), and reaction precursor (*right*) aqueous samples irradiated by UV lamp (*up*) or ambient light (*bottom*). **f** Fluorescent intensity of the prepared SiNPs under different pH values ranging from 1 to 10. **g** Temporal evolution of fluorescent intensity of FITC, CdTe QDs, the as-prepared SiNPs during 120-min UV irradiation. Reprinted with permission from Ref. [31]. Copyright 2011 American Chemical Society

~2.2 nm) SiNPs were facilely achieved via in situ growth under microwave reaction. Of particular significance, this microwave-assisted method holds high promise for large-scale synthesis of fluorescence SiNPs (e.g., only 10 min were required for preparation of 0.1 g SiNPs by using this strategy).

2.2 Silicon Nanowires

SiNWs are regarded as the most important one-dimensional silicon nanostructure, and have obtained giant attentions thus far [49, 50]. Through the past 50-year development, significant progress has been achieved to fabricate SiNWs with controlled morphologies and properties.

As early as 1957, Treuting et al. reported the first preparation of Si whiskers with <111> orientation, or filamentary Si crystals with macroscopic dimensions [51]. Thereafter, Wagner and Ellis performed the illuminating work and established the vapor-liquid-solid (VLS) mechanism of the Si whisker growth [52]. These pioneer studies open exciting avenues for fabrication of SiNWs. The second

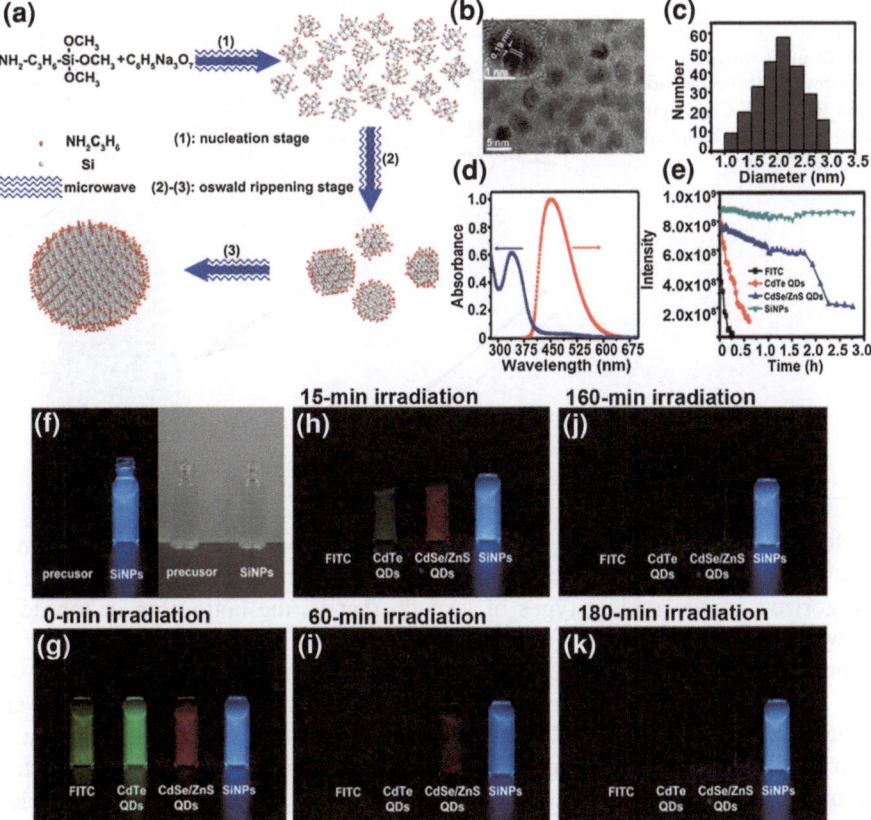

Fig. 2.4 **a** Schematic mode of the "bottom-up" strategy. **b** TEM/HRTEM, **c** size distribution, and **d** absorption and photoluminescence (UV−PL) spectra of the SiNPs prepared through microwave-assisted strategy. Inset in (**b**) presents the enlarged HRTEM image of a single SiNP. **e** Temporal evolution of fluorescence intensity of FITC, CdTe and CdSe/ZnS QDs, and the as-prepared SiNPs under long-term UV irradiation. **f** Photos of reaction precursors and the prepared SiNPs aqueous solution irradiated by UV light (*left*) or ambient light (*right*). **g–k** Pictures of fours aqueous samples (i.e., FITC, CdTe and CdSe/ZnS QDs, and SiNPs), persistently irradiated by UV lamp (365 nm, 450 W) for 180 min. Reprinted with permission from Ref. [34]. Copyright 2013 American Chemical Society

phase in silicon-wire studies were launched in the mid 1990s triggered by development in microelectronics. In 1998, Lieber's [53] and Lee's groups [54] independently reported the strategy of VLS growth assisted by laser ablation, which is especially suitable for large-quantity preparation of single-crystal SiNWs with controllable diameters and lengths (diameter: 6–20 nm, lengths: 1–30 μm). In general, the VLS process involves four steps as follows (Fig. 2.5) [55], silicon precursor is first decomposed using metal catalysts. The liquid alloy made of silicon and metal is formed in the second step. Afterwards, the resultant silicon-metal alloy is diffused with silicon. In the final step, Si supersaturation leads to

Fig. 2.5 Schematic illustration of the VLS growth mechanism. Reprinted with permission from Ref. [55]. Copyright 2010 American Chemical Society

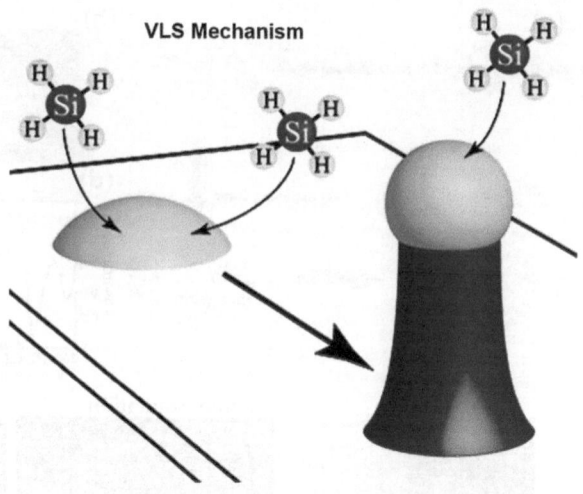

nucleation at the liquid/solid interface and production of SiNWs. Since then, a number of approaches have been developed to synthesize SiNPs, which can be categorized as two typical types of growth, that is, the bottom-up or top-down growth of SiNWs. For the former strategy, SiNWs growth is triggered by an assembly process joining ultrasmall Si atoms, such as metal-catalyzed VLS growth [52, 56, 57], oxide-assisted growth (OAG) [49, 58], supercritical-fluid-based and solution-based growth [59–62], laser ablation [63, 64], and thermal evaporation with catalyst [65, 66]. In contrary, for the top-down approaches (e.g., electron beam lithography (EBL), reaction ion etching (RIE), and metal-catalyzed electroless etching, etc.), large-sized bulk silicon precursor is employed for synthesizing SiNWs via lithography and etching [67–72]. Several representative strategies will be introduced in detail in the following pages.

2.2.1 Chemical Vapor Deposition

Chemical vapor deposition (CVD) is another established strategy especially suitable for preparing vertically aligned SiNWs with high aspect ratio. Metal catalyst (e.g., gold) and gaseous silicon reactants (e.g., silane, SiH_4, disilane, Si_2H_6, silicon dichloride, SiH_2Cl_2, or silicon tetrachloride, $SiCl_4$, etc.) are generally used in the CVD growth. Different reaction temperatures are required to produce SiNWs when the silicon precursors are varied because of their distinct chemical stability. For example, the temperature ranging from 800 to 1000 °C is often required for SiNWs growth by using $SiCl_4$ as silicon precursor. In terms of silane-based silicon precursor, the growth temperature reduces to 400–600 °C [73, 74].

In 2004, Lieber and coworkers employed gold nanoclusers and silane (SiH_4), serving as catalysts and precursors, respectively, to produce ∼3 nm SiNWs with

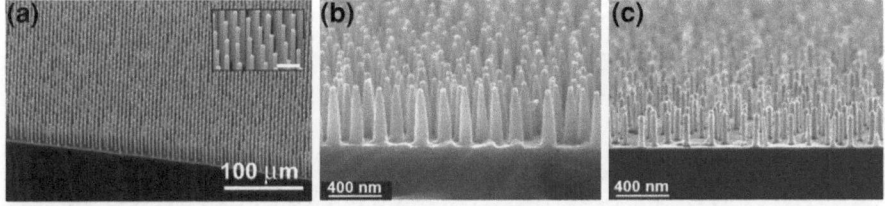

Fig. 2.6 a Tilted SEM overviews of the SiNW arrays with a large area (>1 cm²) using Cu as catalyst. Reprinted with permission from Ref. [79]. Copyright 2007, American Institute of Physics. SEM images of the SiNW arrays prepared at 490 °C (**b**) and 430 °C (**c**). Reprinted with permission from Nature Publishing Group, a division of Macmillan Publishers Ltd: Ref. [82]. Copyright 2006

undetectable amorphous oxide [75]. In this case, hydrogen was utilized since it could passivate the nanowire surface and reduce surface roughness. In 2005, Yang and coworkers used $SiCl_4$ and gold colloids as the precursor and catalyst, respectively, for SiNWs growth under 800–850 °C in the VLS-CVD process [76]. In this method, the gold colloids were capable of defining the size and position of the SiNWs. Notably, the oxide layer on the Si surface could be readily removed by gaseous HCl, which was the byproduct $SiCl_4$ decomposition in H_2, facilitating homoepitaxial growth of SiNWs and production of the SiNWs with clean Si crystal surface. As mentioned above, gold is employed as the most frequently used metal catalyst in the VLS growth, unavoidably leading to gold contamination in the prepared SiNWs [77]. The presence of metallic contaminations may often induce deep-level electronic states of the silicon band gap and degradation of the minority-carrier lifetime, which is adverse to photovoltaic applications. To address this issue, other kinds of metals (e.g., Cu, Al, and Pt, etc.) have been employed as alternative catalysts for synthesizing SiNWs [78–83].

In 2007, using Cu and $SiCl_4$ as catalyst and Si as precursor, respectively, Atwater and coworkers grew large-area (>1 cm²) SiNWs (diameter: ∼1.5 μm, lengths: >75 μm) arrays with vertical orientation [79]. Note that a patterned oxide buffer layer was used in this work to avoid migration of catalysts during the synthetic process, which was favorable for controlling size, position, and uniformity of the SiNWs arrays (Fig. 2.6a). Wang et al. demonstrated that Al could be used as another kind of available catalyst for VLS SiNWs growth, due to similar binary phase diagram of Al-Si [82]. They further revealed that the growth process was probably via CVD-vapor-solid-solid (VSS) mechanism rather than CVD-VLS, because the eutectic temperature of aluminum-silicon binary phase diagram (577 °C) is considerably higher than that employed in the reported method. Moreover, the wires tended to be tapered by using Al as the catalyst. However, the tendency of tapering was found to be reduced by lowering the growth temperature (Fig. 2.6b, c).

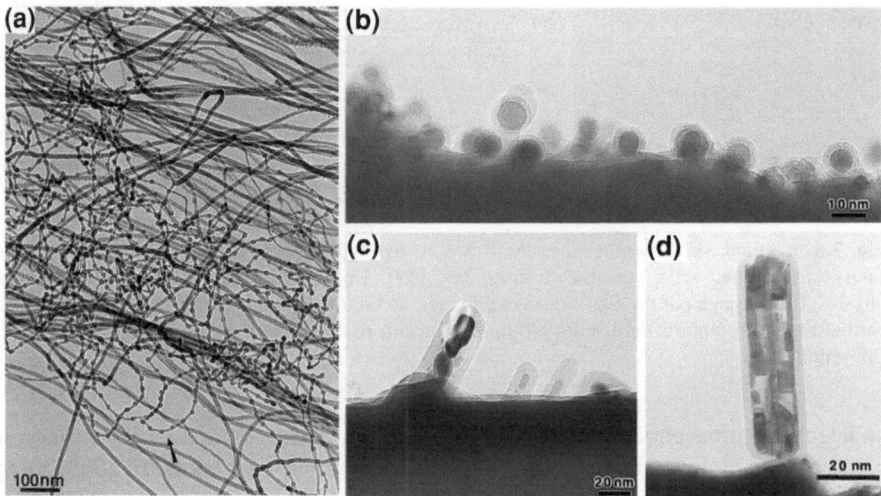

Fig. 2.7 a TEM image of SiNWs prepared through the evaporation method. **b–d** Different nucleation stages of the SiNWs. Reprinted from Ref. [84]. Copyright 1999, with permission from Elsevier

2.2.2 Oxide-Assisted Growth

The OAG method is widely regarded as another popular strategy for fabricating SiNWs. Compared to the metal catalysts-assisted VLS method, oxides are employed in the OAG methods to induce the nucleation and growth of nanowires, producing the SiNWs free of metal contamination. In addition, the OAG method is highly efficacious for producing small-sized SiNWs in large quantities. Moreover, a variety of well-defined silicon structures (e.g., rods, chains, and ribbons, etc.) are readily fabricated using the OAG method [58]. Notwithstanding, the VLS method is more suitable for synthesizing SiNWs with controllable diameters and growth alignment via adjusting sizes and distribution of metal seeds [53, 54, 58, 65, 66]. Consequently, VLS and OAG methods with different shortcomings and advantages are well complementary to each other, and serve as both well-established strategies for synthesis of high-quality SiNWs.

In 1999, Lee and coworkers synthesized bulk-quantity SiNWs by thermal evaporation of a powder mixture of silicon and SiO_2 [84], in which SiNW nucleus containing a polycrystalline Si core was formed at the initial nucleation stage (Fig. 2.7). In their following study, they further fabricated the ultrasmall (diameter: ~ 1 nm) SiNWs via the optimized OAG approach [49], whose surfaces were terminated with hydrogen by a hydrofluoric acid dip. They also observed Si(111) facet and SiH_2 on Si(001) facet in the prepared SiNWs sample via scanning tunneling microscopy (STM) characterization. Besides, the authors found that, in comparison to regular silicon wafer whose surface was easily to be oxidized, the surface of as-prepared SiNWs seemed to be more resistant to oxidation. They further revealed that the electronic energy gaps decreased with increasing SiNW diameter from 3.5 electron volts for 1.3 nm to 1.1 electron volts for 7 nm [49].

2.2.3 Metal-Catalyzed Electroless Etching

Top-down SiNWs growth has recently emerged as an attractive method for preparing SiNWs with desirable flexibility and precision, which is capable of massive and facile production of SiNWs in a low-cost manner.

In 2002, Peng and coworkers introduced a HF-etching-assisted nanoelectrochemical strategy to synthesize wafer-scale aligned SiNWs [69], and further investigated the mechanisms in their following studies [85–88]. The fabrication process is typically divided into two steps: metal catalyst film is first deposited on silicon wafer via electroless metal deposition. Afterwards, the resultant silicon wafer is etched using aqueous HF solution containing oxidizing agents (e.g., H_2O_2, $Fe(NO_3)_3$, or HNO_3, etc.). This metal-catalyzed electroless etching process is basically based on noble metal catalysts-assisted selective oxidization of Si, followed by oxidizing HF solution-induced exclusive etching of silicon at the metal-silicon interface, and eventually yields the SiNWs of controllable lengths (Fig. 2.8) [69–71, 89, 90]. It is worthwhile to point out that this HF-etching method is superbly suited to mild, facile and low-cost fabrication of SiNWs at room temperature and atmospheric pressure, without requiring expensive instruments and reagents.

2.3 Silicon Nanohybrid

Hybrid nanomaterials (nanohybrids) that combine with various types of nanostructures (e.g., nanoparticles, nanowires, and nanotubes) have recently gained great attentions [91–95]. Notably, rational design of the architecture would endow the nanohybrids with desirable features, facilitating investigation of the relationship between nanostructures and electronic/optical/magnetic properties [91–93]. Moreover, to meet the increasing demands of various applications, nanohybrids with multifunctional properties have been fabricated by combinations of various functional nanostructures. In recent years, silicon-based nanohybrids made of metal NPs (e.g., AuNPs, AgNPs, PtNPs, and CdTe QDs, etc.)-decorated SiNWs or SiNPs doped with magnetic materials (e.g., Mn, Fe, Fe_2O_3, Fe_3O_4) have been well developed and utilized for solar cells, catalysts, chemical/biological sensors, cancer therapy, and bioimaging, etc. [10, 96–109].

Basically, noble metal (Pt, Au, and Ag) ions can be facilely reduced by surface-covered Si-H bonds of SiNWs, yielding Pt/Au/Ag NPs-decorated SiNWs. Such resultant metallic nanoparticles decorated-SiNWs have been extensively explored

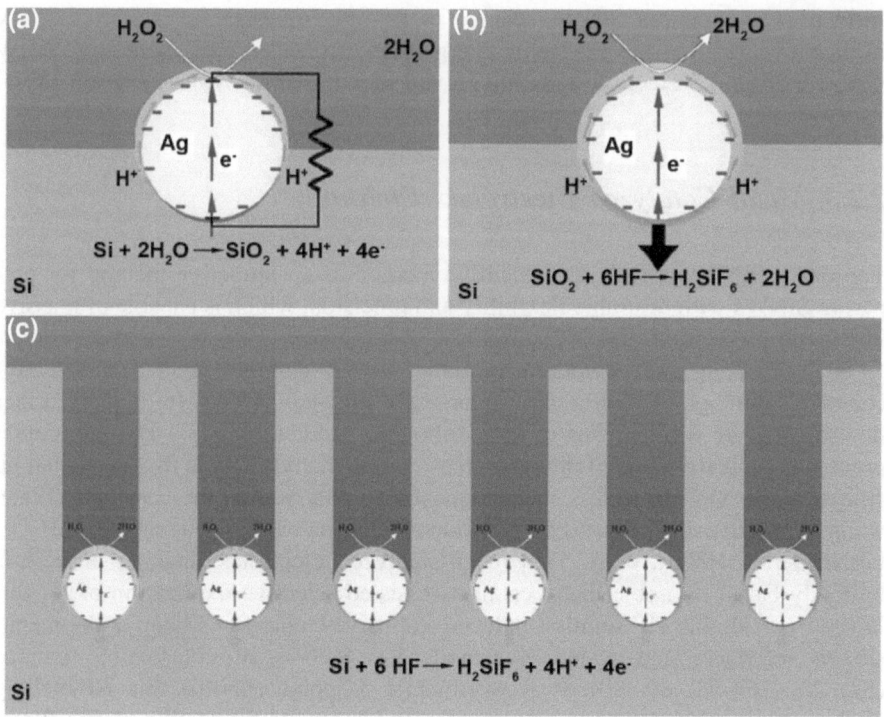

Fig. 2.8 Schematics of Ag particle movement in bulk Si induced by catalysts: **a** A hydrated proton gradient across the Ag particle leads to self-electrophoresis-driven motion. **b** Self-electrophoresis-driven motion of Ag particles. **c** Silicon nanostructures are formed due to tunneling motion of Ag particles in a silicon substrates. Reproduced from Ref. [71] with permission from John Wiley & Sons Inc

for myriad applications (e.g., surface-enhanced Raman scattering (SERS), catalysis, solar cell, and cancer therapy) due to their unique photo/electronic/catalytic properties. For example, in 2011, He and coworkers developed a well-defined SERS platform for DNA detection by using SiNWs decorated with AgNPs (AgNPs@SiNWs). The AgNPs@SiNWs nanohybrid-based biosensor achieved the detection of DNA with a remarkably low concentration (~ 1 fM) due to their very high SERS enhancement factor (up to $\sim 10^{10}$) [100]. Lee and coworkers employed PtNPs-decorated SiNW arrays for fabricating photoelectrochemical solar cells, yielding distinctly improved photoconversion efficiency with high energy conversion efficiency (8.14 %) (Fig. 2.9a, b) [97]. In 2011, He et al. presented a kind of highly fluorescent (QY: ~ 30 %) SiNWs coated with multi-color QDs, capable of cell imaging in long-term manner to take advantage of the excellent anti-photobleaching property of the resultant QDs-decorated SiNWs (Fig. 2.9c, d) [101]. More recently, Su et al. decorated AuNPs on SiNWs surface, and further demonstrated that the prepared AuNPs-decorated SiNWs could produce sufficient heat under near-infrared (NIR) irradiation [103]. In this case, the SiNWs had high

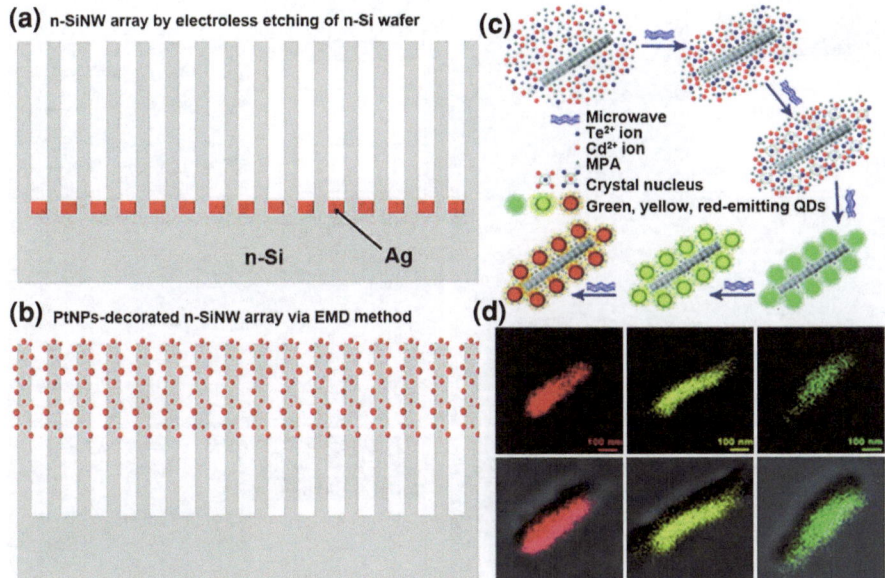

Fig. 2.9 **a** and **b** Schematic illustration of the SiNW arrays whose surface is coated with PtNPs. Reprinted with the permission from Ref. [97]. Copyright 2009 American Chemical Society. **c** Schematics of the preparation of three-color fluorescent QDs-coated SiNWs. **d** Confocal images of the as-prepared three-color (i.e., *green*, *yellow*, and *red*) fluorescent SiNWs. Reproduced from ref. [101] with permission from John Wiley & Sons Inc

absorption at NIR region and converted NIR light into heat, whereas AuNPs coated on surface of SiNWs significantly improve the conversion of light to heat. As a result, such SiNWs-based nanohybrids, served as novel hyperthermia nano-agents, were efficacious for treatment of tumor cells. More examples of silicon nanohybrids-based applications will be introduced in following Chapters (i.e., Chaps. 3, 4 and 5) in a detailed way.

In another aspect, by adding paramagnetism to fluorescent SiNPs, researchers can realize concurrent optical and magnetic (e.g., magnetic resonance imaging (MRI)) detection. Magnetic impurity doping is known as an available approach for offering semiconductor with magnetic properties [110]. In contrast to their single-component counterparts, the resultant doped nanomaterials feature enhanced and multiple functionalities [111]. In 2007, Kauzlarich and coworkers synthesized Mn-doped SiNPs with fluorescent and magnetic properties [105]. In their study, Mn doped Zintl salts ($NaSi_{1-x}Mn_x$, $x = 0.05, 0.1, 0.15$) were reacted with NH_4Br to produce H-terminated nanohybrid. Later, the same group developed another approach to produce fluorescent and paramagnetic Mn-doped SiNPs ($Si_{Mn}NPs$) [107]. In this case, the precursor Mn-doped sodium silicide was first obtained, followed by the reaction with NH_4Br and N,N-dimethylformaminde (DMF) to prepare the hydrophobic $Si_{Mn}NPs$. Finally, the allylamine was added to the reaction mixture to obtain water-soluble $Si_{Mn}NPs$. In 2012, they further

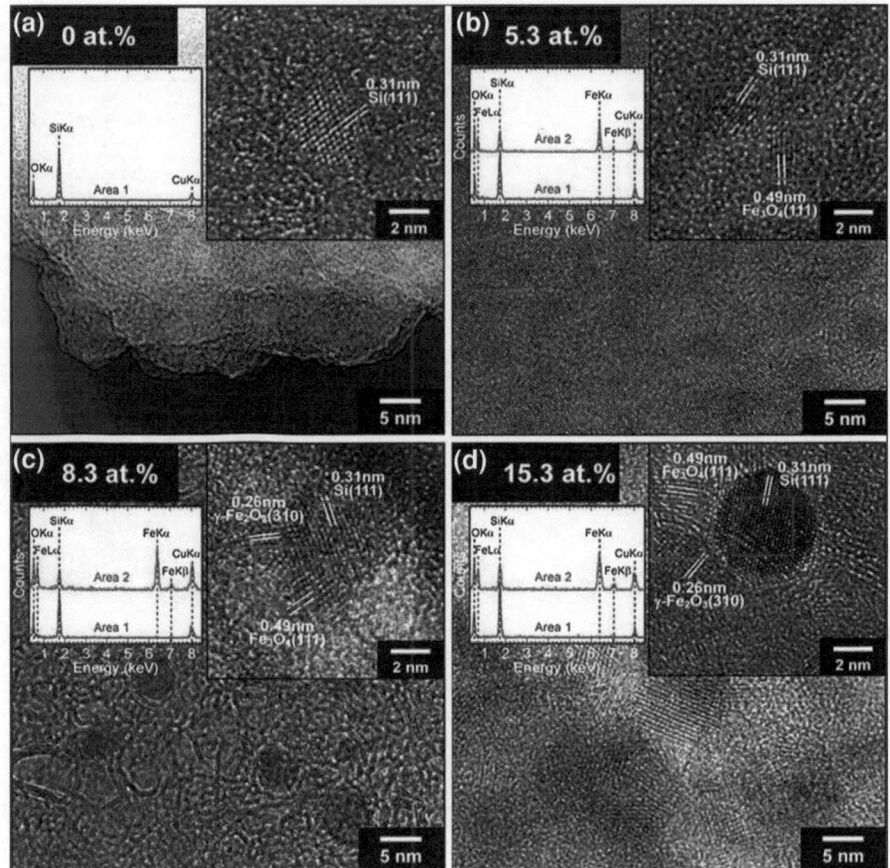

Fig. 2.10 TEM images of the pure SiNPs (**a**) and SiNPs doped with Fe different contents of 5.3 % (**b**), 8.3 % (**c**), and 15.3 % (**d**). Insets in **a–d** show HRTEM and energy dispersive X-ray (EDX) spectra of the mentioned four samples. Reprinted with permission from Ref. [108]. Copyright 2011 American Chemical Society

demonstrated the preparation of allylamine-terminated Fe-doped SiNPs featuring strong fluorescence (QY: ~ 10 %) and significant T_2 contrast, which was suitable for simultaneous fluorescent and magnetic imaging [109].

Another alternative means for fabricating SiNPs with magnetic properties is to encapsulate SiNPs with magnetic NPs, providing higher magnetization with less effect on the luminescence. In 2011, Fukata and coworkers designed the nanohybrids that combined silicon and magnetic iron oxides (e.g., Fe_3O_4 (γ-Fe_2O_3) with size-dependent optical and magnetic behaviors (Fig. 2.10) [108]. For example, the as-prepared nanohybrids with the mean size of 3.0 nm exhibited superparamagnetic behavior and green fluorescence, but showed ferromagnetic behavior without fluorescence when the mean diameter was larger than 5.0 nm.

2.4 Conclusions

In this chapter, we first summarized the representative achievements in the synthesis of fluorescent SiNPs. Several important synthetic strategies (e.g., solution-phase reduction synthesis strategy, electrochemical etching approach, microwave-assisted method, etc.) were reviewed in details. On the basis of which, we further discussed the surface modification of SiNPs and relationship between optical and surface properties of SiNPs. Typically, quantum yield of SiNPs can be distinctly enhanced via proper surface modification (e.g., introducing novel surface ligands). Moreover, we suggested "bottom-up" methods as the highly promising route for large-scale preparation of highly luminescent SiNPs. In the second section, we gave a detailed introduction to three classic methods for synthesizing SiNWs, i.e., CVD, OAG, and electroless etching. We further analyzed advantages and shortcomings of these methods from the viewpoints of synthetic procedures, growth mechanism, catalysts, production cost, etc. We also briefly introduced typical kinds of silicon-based nanohybrids (e.g., metal nanoparticles-decorated SiNWs and magnetic materials-doped SiNPs). These high-quality silicon nanomaterials and their nanohybrids featuring unique optical/electronic/thermal properties afford exciting and new possibilities for myriad biological and biomedical, electronical, catalytic, and energetic applications.

References

1. Michalet X, Pinaud F, Bentolila L, Tsay J, Doose S, Li J, Sundaresan G, Wu A, Gambhir S, Weiss S (2005) Quantum dots for live cells, in vivo imaging, and diagnostics. Science 307(5709):538–544
2. Kostarelos K, Bianco A, Prato M (2009) Promises, facts and challenges for carbon nanotubes in imaging and therapeutics. Nat Nanotechnol 4(10):627–633
3. Riehemann K, Schneider SW, Luger TA, Godin B, Ferrari M, Fuchs H (2009) Nanomedicine-challenge and perspectives. Angew Chem Int Ed 48(5):872–897
4. Hong H, Zhang Y, Sun J, Cai W (2009) Molecular imaging and therapy of cancer with radiolabeled nanoparticles. Nano Today 4(5):399–413
5. Rothenfluh DA, Bermudez H, O'Neil CP, Hubbell JA (2008) Biofunctional polymer nanoparticles for intra-articular targeting and retention in cartilage. Nat Mater 7(3):248–254
6. De M, Ghosh PS, Rotello VM (2008) Applications of nanoparticles in biology. Adv Mater 20(22):4225–4241
7. Pavesi L, Dal Negro L, Mazzoleni C, Franzo G, Priolo F (2000) Optical gain in silicon nanocrystals. Nature 408(6811):440–444
8. Ding Z, Quinn BM, Haram SK, Pell LE, Korgel BA, Bard AJ (2002) Electrochemistry and electrogenerated chemiluminescence from silicon nanocrystal quantum dots. Science 296(5571):1293–1297
9. Grom GF, Lockwood DJ, McCaffrey JP, Labbe HJ, Fauchet PM, White B, Diener J, Kovalev D, Koch F, Tsybeskov L (2000) Ordering and self-organization in nanocrystalline silicon. Nature 407(6802):358–361

10. Allen JE, Hemesath ER, Perea DE, Lensch-Falk JL, LiZ Y, Yin F, Gass MH, Wang P, Bleloch AL, Palmer RE, Lauhon LJ (2008) High-resolution detection of Au catalyst atoms in Si nanowires. Nat Nanotechnol 3(3):168–173
11. Brus L, Szajowski P, Wilson W, Harris T, Schuppler S, Citrin P (1995) Electronic spectroscopy and photophysics of Si nanocrystals: relationship to bulk c-Si and porous Si. J Am Chem Soc 117(10):2915–2922
12. Wilson WL, Szajowski P, Brus L (1993) Quantum confinement in size-selected, surface-oxidized silicon nanocrystals. Science 262:1242–1244
13. Park N-M, Choi C-J, Seong T-Y, Park S-J (2001) Quantum confinement in amorphous silicon quantum dots embedded in silicon nitride. Phys Rev Lett 86(7):1355–1357
14. Canham LT (1990) Silicon quantum wire array fabrication by electrochemical and chemical dissolution of wafers. Appl Phys Lett 57(10):1046–1048
15. Cullis A, Canham L (1991) Visible light emission due to quantum size effects in highly porous crystalline silicon. Nature 353:335–338
16. Yang C-S, Bley RA, Kauzlarich SM, Lee HW, Delgado GR (1999) Synthesis of alkyl-terminated silicon nanoclusters by a solution route. J Am Chem Soc 121(22):5191–5195
17. Baldwin RK, Pettigrew KA, Ratai E, Augustine MP, Kauzlarich SM (2002) Solution reduction synthesis of surface stabilized silicon nanoparticles. Chem Commun 17:1822–1823
18. Tilley RD, Yamamoto K (2006) The microemulsion synthesis of hydrophobic and hydrophilic silicon nanocrystals. Adv Mater 18(15):2053–2056
19. Shiohara A, Hanada S, Prabakar S, Fujioka K, Lim TH, Yamamoto K, Northcote PT, Tilley RD (2010) Chemical reactions on surface molecules attached to silicon quantum dots. J Am Chem Soc 132(1):248–253
20. Arul Dhas N, Raj CP, Gedanken A (1998) Preparation of luminescent silicon nanoparticles: a novel sonochemical approach. Chem Mater 10(11):3278–3281
21. Heintz AS, Fink MJ, Mitchell BS (2007) Mechanochemical synthesis of blue luminescent alkyl/alkenyl-passivated silicon nanoparticles. Adv Mater 19(22):3984–3988
22. Riabinina D, Durand C, Chaker M, Rosei F (2006) Photoluminescent silicon nanocrystals synthesized by reactive laser ablation. Appl Phys Lett 88(7):073105
23. Jurbergs D, Rogojina E, Mangolini L, Kortshagen U (2006) Silicon nanocrystals with ensemble quantum yields exceeding 60 %. Appl Phys Lett 88(23):233116
24. Mangolini L, Thimsen E, Kortshagen U (2005) High-yield plasma synthesis of luminescent silicon nanocrystals. Nano Lett 5(4):655–659
25. Mangolini L, Kortshagen U (2007) Plasma-assisted synthesis of silicon nanocrystal inks. Adv Mater 19(18):2513–2519
26. Kim NY, Laibinis PE (1997) Thermal derivatization of porous silicon with alcohols. J Am Chem Soc 119(9):2297–2298
27. Kang Z, Tsang CHA, Zhang Z, Zhang M, Wong N-b, Zapien JA, Shan Y, Lee S-T (2007) A polyoxometalate-assisted electrochemical method for silicon nanostructures preparation: from quantum dots to nanowires. J Am Chem Soc 129(17):5326–5327
28. Kang Z, Liu Y, Tsang CHA, Ma DDD, Fan X, Wong NB, Lee S-T (2009) Water-soluble silicon quantum dots with wavelength-tunable photoluminescence. Adv Mater 21(6):661–664
29. He Y, Kang ZH, Li QS, Tsang CHA, Fan CH, Lee S-T (2009) Ultrastable, highly Fluorescent, and water-dispersed silicon-based nanospheres as cellular probes. Angew Chem Int Ed 48:128–132
30. He Y, Su Y, Yang X, Kang Z, Xu T, Zhang R, Fan C, Lee S-T (2009) Photo and pH stable, highly-luminescent silicon nanospheres and their bioconjugates for immunofluorescent cell imaging. J Am Chem Soc 131(12):4434–4438
31. He Y, Zhong Y, Peng F, Wei X, Su Y, Lu Y, Su S, Gu W, Liao L, Lee S-T (2011) One-pot microwave synthesis of water-dispersible, ultraphoto-and pH-stable, and highly fluorescent silicon quantum dots. J Am Chem Soc 133(36):14192–14195

32. Atkins TM, Thibert A, Larsen DS, Dey S, Browning ND, Kauzlarich SM (2011) Femtosecond ligand/core dynamics of microwave-assisted synthesized silicon quantum dots in aqueous solution. J Am Chem Soc 133(51):20664–20667
33. Zhong Y, Peng F, Wei X, Zhou Y, Wang J, Jiang X, Su Y, Su S, Lee S-T, He Y (2012) Microwave-assisted synthesis of biofunctional and fluorescent silicon nanoparticles using proteins as hydrophilic ligands. Angew Chem Int Ed 51(34):8485–8489
34. Zhong Y, Peng F, Bao F, Wang S, Ji X, Yang L, Su Y, Lee S-T, He Y (2013) Large-scale aqueous synthesis of fluorescent and biocompatible silicon nanoparticles and their use as highly photostable biological probes. J Am Chem Soc 135(22):8350–8356
35. Li QS, Zhang RQ, Niehaus TA, Frauenheim T, Lee S-T (2007) Theoretical studies on optical and electronic properties of propionic-acid-terminated silicon quantum dots. J Chem Theory Comput 3(4):1518–1526
36. Li Q, Zhang R, Lee S, Niehaus TA, Frauenheim T (2008) Amine-capped silicon quantum dots. Appl Phys Lett 92(5):053107
37. Wang X, Zhang R, Niehaus TA, Frauenheim T (2007) Excited state properties of allylamine-capped silicon quantum dots. J Phys Chem C 111(6):2394–2400
38. Puzder A, Williamson AJ, Grossman JC, Galli G (2003) Computational studies of the optical emission of silicon nanocrystals. J Am Chem Soc 125(9):2786–2791
39. Zhou Z, Brus L, Friesner R (2003) Electronic structure and luminescence of 1.1-and 1.4-nm silicon nanocrystals: oxide shell versus hydrogen passivation. Nano Lett 3(2):163–167
40. Li Q, He Y, Chang J, Wang L, Chen H, Tan Y-W, Wang H, Shao Z (2013) Surface-modified silicon nanoparticles with ultrabright photoluminescence and single-exponential decay for nanoscale fluorescence lifetime imaging of temperature. J Am Chem Soc 135(40):14924–14927
41. Song S, Qin Y, He Y, Huang Q, Fan C, Chen H-Y (2010) Functional nanoprobes for ultrasensitive detection of biomolecules. Chem Soc Rev 39(11):4234–4243
42. He Y, Fan C, Lee S-T (2010) Silicon nanostructures for bioapplications. Nano Today 5(4):282–295
43. Li Z, Ruckenstein E (2004) Water-soluble poly (acrylic acid) grafted luminescent silicon nanoparticles and their use as fluorescent biological staining labels. Nano Lett 4(8):1463–1467
44. Sato S, Swihart MT (2006) Propionic-acid-terminated silicon nanoparticles: synthesis and optical characterization. Chem Mater 18(17):4083–4088
45. Warner JH, Hoshino A, Yamamoto K, Tilley RD (2005) Water-soluble photoluminescent silicon quantum dots. Angew Chem Int Ed 44(29):4550–4554
46. Erogbogbo F, Yong K-T, Roy I, Xu G, Prasad PN, Swihart MT (2008) Biocompatible luminescent silicon quantum dots for imaging of cancer cells. ACS Nano 2(5):873–878
47. Choi HS, Liu W, Misra P, Tanaka E, Zimmer JP, Ipe BI, Bawendi MG, Frangioni JV (2007) Renal clearance of quantum dots. Nat Biotechnol 25(10):1165–1170
48. Choi HS, Liu W, Liu F, Nasr K, Misra P, Bawendi MG, Frangioni JV (2010) Design considerations for tumour-targeted nanoparticles. Nat Nanotechnol 5(1):42–47
49. Ma D, Lee C, Au F, Tong S, Lee S-T (2003) Small-diameter silicon nanowire surfaces. Science 299(5614):1874–1877
50. Schmidt V, Wittemann JV, Senz S, Gösele U (2009) Silicon nanowires: A review on aspects of their growth and their electrical properties. Adv Mater 21(25–26):2681–2702
51. Treuting RG, Arnold SM (1957) Orientation habits of metal whiskers. Acta Metall 5(10):598
52. Wagner R, Ellis W (1964) Vapor-liquid-solid mechanism of single crystal growth. Appl Phys Lett 4(5):89–90
53. Morales AM, Lieber CM (1998) A laser ablation method for the synthesis of crystalline semiconductor nanowires. Science 279(5348):208–211
54. Zhang YF, Tang YH, Wang N, Yu DP, Lee CS, Bello I, Lee S-T (1998) Silicon nanowires prepared by laser ablation at high temperature. Appl Phys Lett 72(15):1835–1837

55. Schmidt V, Wittemann J, Gosele U (2010) Growth, thermodynamics, and electrical properties of silicon nanowires. Chem Rev 110(1):361–388
56. Wu Y, Yang P (2001) Direct observation of vapor-liquid-solid nanowire growth. J Am Chem Soc 123(13):3165–3166
57. Putnam MC, Filler MA, Kayes BM, Kelzenberg MD, Guan Y, Lewis NS, Eiler JM, Atwater HA (2008) Secondary ion mass spectrometry of vapor-liquid-solid grown, Au-catalyzed Si wires. Nano Lett 8(10):3109–3113
58. Zhang RQ, Lifshitz Y, Lee S-T (2003) Oxide-assisted growth of semiconducting nanowires. Adv Mater 15(7–8):635–640
59. Holmes JD, Johnston KP, Doty RC, Korgel BA (2000) Control of thickness and orientation of solution-grown silicon nanowires. Science 287(5457):1471–1473
60. Lu X, Hanrath T, Johnston KP, Korgel BA (2002) Growth of single crystal silicon nanowires in supercritical solution from tethered gold particles on a silicon substrate. Nano Lett 3(1):93–99
61. Heitsch AT, Fanfair DD, Tuan H-Y, Korgel BA (2008) Solution-liquid-solid (SLS) growth of silicon nanowires. J Am Chem Soc 130(16):5436–5437
62. Hanrath T, Korgel BA (2003) Supercritical fluid-liquid-solid (SFLS) synthesis of Si and Ge nanowires seeded by colloidal metal nanocrystals. Adv Mater 15(5):437–440
63. Zhang YF, Tang YH, Peng HY, Wang N, Lee CS, Bello I, Lee S-T (1999) Diameter modification of silicon nanowires by ambient gas. Appl Phys Lett 75(13):1842–1844
64. Tang YH, Zhang YF, Wang N, Lee CS, Han XD, Bello I, Lee S-T (1999) Morphology of Si nanowires synthesized by high-temperature laser ablation. J Appl Phys 85(11):7981–7983
65. Shi WS, Peng HY, Zheng YF, Wang N, Shang NG, Pan ZW, Lee CS, Lee S-T (2000) Synthesis of large areas of highly oriented, very long silicon nanowires. Adv Mater 12(18):1343–1345
66. Pan H, Lim S, Poh C, Sun H, Wu X, Feng Y, Lin J (2005) Growth of Si nanowires by thermal evaporation. Nanotechnology 16(4):417
67. Juhasz R, Elfström N, Linnros J (2005) Controlled fabrication of silicon nanowires by electron beam lithography and electrochemical size reduction. Nano Lett 5(2):275–280
68. Tong HD, Chen S, van der Wiel WG, Carlen ET, van den Berg A (2009) Novel top-down wafer-scale fabrication of single crystal silicon nanowires. Nano Lett 9(3):1015–1022
69. Peng KQ, Yan YJ, Gao SP, Zhu J (2002) Synthesis of large-area silicon nanowire arrays via self-assembling nanoelectrochemistry. Adv Mater 14(16):1164–1167
70. Peng K, Wu Y, Fang H, Zhong X, Xu Y, Zhu J (2005) Uniform, axial-orientation alignment of one-dimensional single-crystal silicon nanostructure arrays. Angew Chem Int Ed 44(18):2737–2742
71. Peng K, Lu A, Zhang R, Lee S-T (2008) Motility of metal nanoparticles in silicon and induced anisotropic silicon etching. Adv Funct Mater 18(19):3026–3035
72. Hsu C-M, Connor ST, Tang MX, Cui Y (2008) Wafer-scale silicon nanopillars and nanocones by Langmuir-Blodgett assembly and etching. Appl Phys Lett 93(13):133109
73. Lew K-K, Redwing JM (2003) Growth characteristics of silicon nanowires synthesized by vapor–liquid–solid growth in nanoporous alumina templates. J Cryst Growth 254(1–2):14–22
74. Chung S-W, Yu J-Y, Heath JR (2000) Silicon nanowire devices. Appl Phys Lett 76(15):2068–2070
75. Wu Y, Cui Y, Huynh L, Barrelet CJ, Bell DC, Lieber CM (2004) Controlled growth and structures of molecular-scale silicon nanowires. Nano Lett 4(3):433–436
76. Hochbaum AI, Fan R, He R, Yang P (2005) Controlled growth of Si nanowire arrays for device integration. Nano Lett 5(3):457–460
77. Hannon J, Kodambaka S, Ross F, Tromp R (2006) The influence of the surface migration of gold on the growth of silicon nanowires. Nature 440(7080):69–71
78. Garnett EC, Liang W, Yang P (2007) Growth and electrical characteristics of platinum-nanoparticle-catalyzed silicon nanowires. Adv Mater 19(19):2946–2950

79. Kayes BM, Filler MA, Putnam MC, Kelzenberg MD, Lewis NS, Atwater HA (2007) Growth of vertically aligned Si wire arrays over large areas (>1 cm^2) with Au and Cu catalysts. Appl Phys Lett 91(10):103110
80. Renard VT, Jublot M, Gergaud P, Cherns P, Rouchon D, Chabli A, Jousseaume V (2009) Catalyst preparation for CMOS-compatible silicon nanowire synthesis. Nat Nanotechnol 4(10):654–657
81. Putnam MC, Turner-Evans DB, Kelzenberg MD, Boettcher SW, Lewis NS, Atwater HA (2009) 10 µm minority-carrier diffusion lengths in Si wires synthesized by Cu-catalyzed vapor-liquid-solid growth. Appl Phys Lett 95:163116
82. Wang Y, Schmidt V, Senz S, Gösele U (2006) Epitaxial growth of silicon nanowires using an aluminium catalyst. Nat Nanotechnol 1(3):186–189
83. Wacaser BA, Reuter MC, Khayyat MM, Wen C-Y, Haight R, Guha S, Ross FM (2009) Growth system, structure, and doping of aluminum-seeded epitaxial silicon nanowires. Nano Lett 9(9):3296–3301
84. Wang N, Tang YH, Zhang YF, Lee CS, Bello I, Lee S-T (1999) Si nanowires grown from silicon oxide. Chem Phys Lett 299(2):237–242
85. Peng K, Zhu J (2003) Simultaneous gold deposition and formation of silicon nanowire arrays. J Electroanal Chem 558:35–39
86. Peng K, Zhu J (2004) Morphological selection of electroless metal deposits on silicon in aqueous fluoride solution. Electrochim Acta 49(16):2563–2568
87. Peng K, Fang H, Hu J, Wu Y, Zhu J, Yan Y, Lee S (2006) Metal-particle-induced, highly localized site-specific etching of Si and formation of single-crystalline Si nanowires in aqueous fluoride solution. Chem Eur J 12(30):7942–7947
88. Peng K, Hu J, Yan Y, Wu Y, Fang H, Xu Y, Lee S, Zhu J (2006) Fabrication of single-crystalline silicon nanowires by scratching a silicon surface with catalytic metal particles. Adv Funct Mater 16(3):387–394
89. Peng K, Xu Y, Wu Y, Yan Y, Lee S-T, Zhu J (2005) Aligned single-crystalline Si nanowire arrays for photovoltaic applications. Small 1(11):1062–1067
90. Peng K, Wang X, Lee S-T (2008) Silicon nanowire array photoelectrochemical solar cells. Appl Phys Lett 92(16):163103
91. Lam HY, Zangmeister CD, Kushmerick JG (2007) Origin of discrepancies in inelastic electron tunneling spectra of molecular junctions. Phys Rev Lett 98(20):206803
92. Mann S (2009) Self-assembly and transformation of hybrid nano-objects and nanostructures under equilibrium and non-equilibrium conditions. Nat Mater 8(10):781–792
93. Macdonald JE, Bar Sadan M, Houben L, Popov I, Banin U (2010) Hybrid nanoscale inorganic cages. Nat Mater 9(10):810–815
94. Peng X, Chen J, Misewich JA, Wong SS (2009) Carbon nanotube-nanocrystal heterostructures. Chem Soc Rev 38(4):1076–1098
95. Wada A, Tamaru S, Ikeda M, Hamachi I (2009) MCM-enzyme-supramolecular hydrogel hybrid as a fluorescence sensing material for polyanions of biological significance. J Am Chem Soc 131(14):5321–5330
96. Su S, Wei X, Zhong Y, Guo Y, Su Y, Huang Q, Lee S-T, Fan C, He Y (2012) Silicon nanowire-based molecular beacons for high-sensitivity and sequence-specific DNA multiplexed analysis. ACS Nano 6(3):2582–2590
97. Peng K-Q, Wang X, Wu X-L, Lee S-T (2009) Platinum nanoparticle decorated silicon nanowires for efficient solar energy conversion. Nano Lett 9(11):3704–3709
98. Peng K-Q, Lee S-T (2011) Silicon nanowires for photovoltaic solar energy conversion. Adv Mater 23(2):198–215
99. Peng Z, Hu H, Utama MIB, Wong LM, Ghosh K, Chen R, Wang S, Shen Z, Xiong Q (2010) Heteroepitaxial decoration of Ag nanoparticles on Si nanowires: a case study on Raman scattering and mapping. Nano Lett 10(10):3940–3947
100. He Y, Su S, Xu T, Zhong Y, Zapien JA, Li J, Fan C, Lee S-T (2011) Silicon nanowires-based highly-efficient SERS-active platform for ultrasensitive DNA detection. Nano Today 6(2):122–130

101. He Y, Zhong Y, Peng F, Wei X, Su Y, Su S, Gu W, Liao L, Lee S-T (2011) Highly luminescent water-dispersible silicon nanowires for long-term immunofluorescent cellular imaging. Angew Chem Int Ed 50:3080–3083
102. Lv M, Su S, He Y, Huang Q, Hu W, Li D, Fan C, Lee S-T (2010) Long-term antimicrobial effect of silicon nanowires decorated with silver nanoparticles. Adv Mater 22(48):5463–5467
103. Su Y, Wei X, Peng F, Zhong Y, Lu Y, Su S, Xu T, Lee S-T, He Y (2012) Gold nanoparticles-decorated silicon nanowires as highly efficient near-infrared hyperthermia agents for cancer cells destruction. Nano Lett 12(4):1845–1850
104. Park G-S, Kwon H, Kwak DW, Park SY, Kim M, Lee J-H, Han H, Heo S, Li XS, Lee JH (2012) Full surface embedding of gold clusters on silicon nanowires for efficient capture and photothermal therapy of circulating tumor cells. Nano Lett 12(3):1638–1642
105. Zhang X, Brynda M, Britt RD, Carroll EC, Larsen DS, Louie AY, Kauzlarich SM (2007) Synthesis and characterization of manganese-doped silicon nanoparticles: bifunctional paramagnetic-optical nanomaterial. J Am Chem Soc 129(35):10668–10669
106. Erogbogbo F, Yong K-T, Hu R, Law W-C, Ding H, Chang C-W, Prasad PN, Swihart MT (2010) Biocompatible magnetofluorescent probes: luminescent silicon quantum dots coupled with superparamagnetic iron (III) oxide. ACS Nano 4(9):5131–5138
107. Tu C, Ma X, Pantazis P, Kauzlarich SM, Louie AY (2010) Paramagnetic, silicon quantum dots for magnetic resonance and two-photon imaging of macrophages. J Am Chem Soc 132(6):2016–2023
108. Sato K, Yokosuka S, Takigami Y, Hirakuri K, Fujioka K, Manome Y, Sukegawa H, Iwai H, Fukata N (2011) Size-tunable silicon/iron oxide hybrid nanoparticles with fluorescence, superparamagnetism, and biocompatibility. J Am Chem Soc 133(46):18626–18633
109. Singh MP, Atkins TM, Muthuswamy E, Kamali S, Tu C, Louie AY, Kauzlarich SM (2012) Development of iron-doped silicon nanoparticles as bimodal imaging agents. ACS Nano 6(6):5596–5604
110. Shi W, Zeng H, Sahoo Y, Ohulchanskyy TY, Ding Y, Wang ZL, Swihart M, Prasad PN (2006) A general approach to binary and ternary hybrid nanocrystals. Nano Lett 6(4):875–881
111. Wang S, Jarrett BR, Kauzlarich SM, Louie AY (2007) Core/shell quantum dots with high relaxivity and photoluminescence for multimodality imaging. J Am Chem Soc 129(13):3848–3856

Chapter 3
Silicon-Based Platform for Biosensing Applications

Abstract Development of high-performance biosensors vastly facilitates the analysis and detection of various biological species, including nucleic acids, protein, cell, etc. Functional nanomaterials (e.g., silver/gold nanoparticles, carbon nanotubes, graphene, silicon nanowires, etc) serve as new platform for design of nano-biosensors featuring high sensitivity and specificity. Taking advantage of the attractive merits of silicon nanowires (SiNWs) (e.g., unique electronic/optical properties, huge surface-to-volume rations, surface tailorability, fast response and good reproducibility, and compatibility with conventional silicon technology), SiNWs have been widely employed for constructing various kinds of electro-chemical and optical biosensors, enabling ultrasensitive, specific, and reproducible detection of DNA and protein. We introduce a number of typical SiNWs-based biosensors (e.g., field-effect transistor (FET), amperometric-, surface-enhanced Raman scattering (SERS), and fluorescence-based biosensors) in this chapter, aiming to summarize the representative progresses of this research field in recent years. These kinds of high-quality silicon-based sensors show potentially great promise for myriad practical applications, such as medical diagnosis, food safety, drug security, environment monitoring, as well as anti-bioterrorism and so forth.

Keywords Silicon nanowires · Biosensor · Field effect transistor · Surface-enhanced Raman scattering (SERS) · DNA and protein detection · Sensitivity and specificity

Biological/chemical analysis and detection are of essential importance for disease diagnosis, drug discovery, food safety, environmental protection, and anti-biot-errorism, etc. While a large number of bioassay kits have been available on the market, there are ever-growing efforts to develop new detection methods with ultrahigh sensitivity and specificity, to meet the increasing demands of biosensing applications. Parallel to research efforts devoted to such high-end requirements, much recent interest has been directed toward the development of low-cost and portable biosensors, which possess comparable high sensitivity to conventional instruments, but with greatly reduced cost/mass/power requirements. Nanostructures (e.g., nanoparticles [1–9], nanotubes [10–14], and nanowires [15–18], etc)

Y. He and Y. Su, *Silicon Nano-biotechnology*, SpringerBriefs in Molecular Science, DOI: 10.1007/978-3-642-54668-6_3, © The Author(s) 2014

featuring unique optical/magnetic/electrical properties, provide new opportunities for this key task. Among them, SiNWs, as the most important one-dimensional silicon nanostructures, are of particular interest, due to their many unique merits including favorable biocompatibility, facile surface modification, large surface-to-volume ratios, rapid response, and adaptable reproducibility [19–22]. The above merits have motivated intensive investigation of SiNWs for various biological sensing applications. In this chapter, we aim to summarize the recent achievement of designing SiNWs-based electrochemical and optical sensors for sensitive and specific detection of different biological targets (e.g., DNA, protein, cell, etc).

In Sect. 3.1, we give a detailed description of SiNWs-based electrochemical biosensors, including field-effect transistor (FET) and amperometric-based biosensors for DNA and protein detection. Fabrication of optical biosensors (e.g., surface-enhanced Raman scattering (SERS) or fluorescence-based sensors) assisted by using silicon materials is introduced in Sect. 3.2. We further discuss opportunities and challenges for silicon nanostructures-based biosensors in the final section, i.e., Sect. 3.3.

3.1 SiNWs-Based Electrochemical Biosensor

An electrochemical biosensor, as defined by the International Union of Pure and Applied Chemistry (IUPAC), is a self-contained integrated device enabling specific analysis and detection assisted by biological recognition elements (biochemical receptors) in direct spatial contact with a transduction element [23–25]. The electrochemical biosensors enable facile and quick analysis and detection of biological species. In the following pages, we will mainly introduce two typical kinds of SiNWs-based biosensors, i.e., FET and amperometric-based biosensor, in a detailed way.

3.1.1 SiNWs-Based Field-Effect Transistor

The FET is highly efficacious for amplifying feeble signals, in which current flows along a semiconductor path (so-called channel) and two electrodes (so-called source and drain) are connected using the semiconductor [26]. Typically, in a p-type semiconductor, the conductance of semiconductor is increased or decreased through applying a negative or positive gate voltage. More importantly, FET is especially suitable for analyzing alteration of electronic signals induced by biological or chemical species. Currently, a variety of nanomaterials have been employed for constructing FET with high sensitivity. For example, Dai and coworkers developed carbon nanotube-based FET for detection of protein or protein-protein interaction in wide-ranging concentrations from 100 pM to 100 nM [13]. Chen et al. presented a kind of GaN nanowires based extended-gate

FET biosensor capable of specific and sensitive (10^{-18} M) DNA sequence identification [27].

Mature SiNWs fabrication techniques significantly facilitate development of SiNWs-based FETs (SiNW-FET). As mentioned in Chap. 2, sizes [21], shapes [28], and dopants [29] of SiNWs can be precisely tailored with good reproducibility via various established synthetic strategies, greatly facilitating the fabrication of SiNWs-based FET with high sensitivity and multiplex capability [30]. Till now, various SiNW-FET nanosensor devices have been designed for detecting a variety of species, including nucleic acids [31–41], proteins [15, 42–47], chemical species and metal ions [15, 48–51], single virus particles [52], cells [53], and even interaction between molecules [54–56] and cell's life activities [57–59]. Moreover, during the past several years, great efforts have been paid to the comprehensive understanding of the filed effect in SiNWs sensors, and advance widespread adoption of the technology for routine use. It has been known that several factors may affect the field effect, such as surface chemistry [33, 35], SiNWs size [60], Debye screening [61], charge layer distance [62], SiNWs shape [37, 38], and so on. In order to gain a better understanding, several typical examples of biosensing applications using SiNW-FETs devices will be described as follows.

3.1.1.1 SiNWs-Based FET for Nucleic Acids Detection

Most genetic information (e.g., growth procedures of all cells and many viruses) is known to be stored in nucleic acids. In recent years, great strides have been made in the development of sensitive, specific, reliable, and rapid DNA biosensors, offering new opportunities for wide-ranging applications including molecular biology research and clinic disease diagnosis [63–66]. While well-established polymerase chain reaction (PCR)-based assays are highly sensitive, they are nevertheless difficult to profile genes of extremely low concentrations in a reliable manner [67]. Remarkable advancement of nanotechnology provides exciting feasibility for sensitive and specific detection of nucleic acid [10–16]. Of particular note, SiNW-FET devices were recently found to be superbly suited to real-time and label-free detection of various biological species with ultrahigh sensitivity.

Early in 2004, Li and coworkers presented a class of SiNWs field-effect devices for detecting single-stranded DNA [32]. Typically, SiNWs surface was first fabricated with single stranded (ss) DNA probes. As a result, hybridization between the capture DNA and complementary ss-DNA (target) led to the change of the SiNWs conductance. In their experiment, for a 12-mer oligonucleotide probe, low-concentration target DNA (25 pM) could be sensitively detected with a high signal/noise ratio larger than 6, which was in sharp contrast to feeble signals of 12-mers with one-base [32]. Meanwhile, Lieber et al. employed peptide nucleic acid (PNAs) as the recognition sequence for facilely recognizing target DNA using the SiNWs-based FET (Fig. 3.1) [31]. Therein, the surfaces of the p-type SiNWs devices were modified with PNA receptors to discriminate wiled type, allowing specific and sensitive detection of wild-type sequences with low concentrations (10 fM) [31].

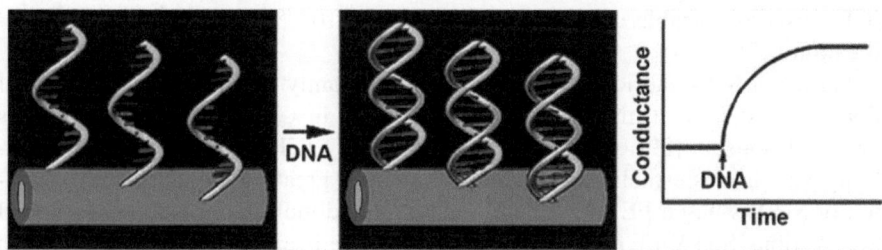

Fig. 3.1 Schematic illustration of the PNA-functionalized SiNW-based FET sensing device for DNA detection based on observing the change of conductance response from the SiNW. Reprinted with permission from Ref. [31]. Copyright 2004 American Chemical Society

Later, Gao and coworkers fabricated SiNW arrays-based DNA biosensors with a detection limit of 10 fM, where the linear detection of target DNA over a wide dynamic range was achieved [34]. Zhang et al. developed a SiNWs device-assisted label-free assay for microRNA (miRNA, recognized as a key player in gene regulation, are small RNA oligonucleotides which bind to mRNAs) detection [36]. In this work, PNAs were immobilized on the surface of the SiNWs device to recognize miRNA, with a low detection limit of 1 fM in the optimized assay. Moreover, the SiNWs device was further employed for detecting miRNA in total RNA extracted from cells, suggesting new possibilities for sensitive and specific detection of miRNA as a biomarker in early diagnosis of cancer [36].

On the other hand, many efforts have been devoted to better understand the mechanism and improving the performance of the filed effect in SiNWs sensors. For example, Heath et al. compared the functionalization chemistries of different types of SiNWs surfaces, revealing that SiNWs without the native oxide exhibited superior FET characteristics and improved sensitivity to ssDNA detection (Fig. 3.2) [33]. They further employed the SiNWs-based sensors for real-time and label-free detection in physiological solution [33]. Zhang and coworkers finely controlled hybridization sites of DNA to PNA preimmobilized on SiNWs surface, realizing controllable distance of a charge layer. On the basis of which, they revealed that the SiNWs-based FET performance was largely relied on the location of the charge layer (Fig. 3.3), which was further proved by theoretical analysis [62]. Recently, Fan and coworkers designed another kind of high-performance SiNWs-based FET devices [37, 38], enabling sensitive and specific detection of DNA with a low concentration of 1 fM (Fig. 3.4). The resultant sensor was also suitable for multiplexing detection of two kinds of pathogenic strain virus DNA sequences (i.e., H1N1 and H5N1) [37]. The same group further developed rolling circle amplification (RCA)-assisted SiNW-FET for DNA detection with high sensitivity and specificity [41]. In this case, SiNWs surface was first immobilized with probe DNA stands, and then further hybridized with complementary target DNA and RCA primer (Fig. 3.5). The RCA reaction led to the creation of a long ssDNA product and the significance enhancement of electronic responses of

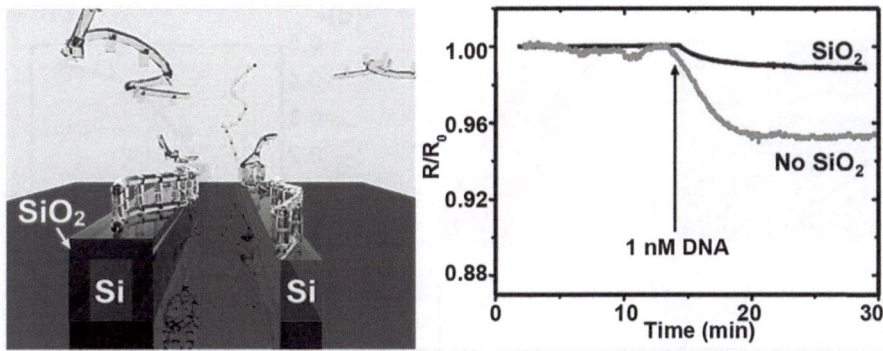

Fig. 3.2 *Left* schematic representation of pure SiNWs and the SiO₂-coated SiNWs. *Right* comparison of conductance response from the pure SiNWs and the SiO₂-coated SiNWs coating with 1 nM DNA. Reprinted with permission from Ref. [33]. Copyright 2006 American Chemical Society

Fig. 3.3 a Schematic illustration of difference of the field effect of the SiNW sensing device resulted from different hybridization sites of target DNA to PNA. **b** Various hybridization sites-induced distinct resistance change of the SiNW sensors. Reprinted with permission from Ref. [62]. Copyright 2008 American Chemical Society

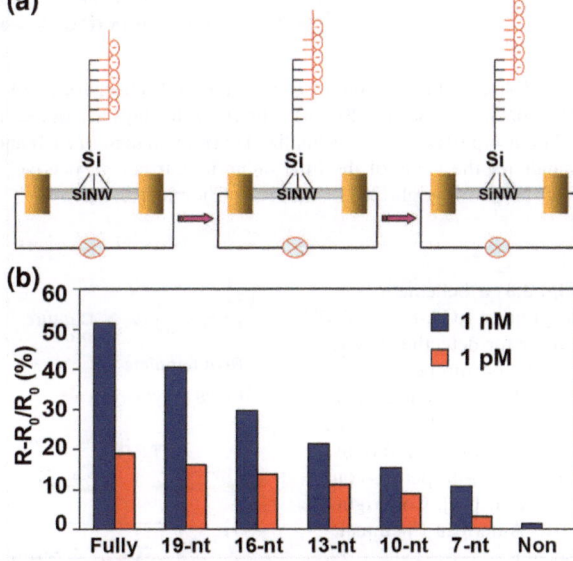

SiNWs, allowing sensitive detection of 1 fM DNA with a high signal-to-noise ratio (>20) [41].

3.1.1.2 SiNW-FET for Proteins Detection

The first example of SiNW-FET-based protein detection was reported by Lieber's group in 2001 by employing single-crystal boron-doped (p-type) SiNWs devices,

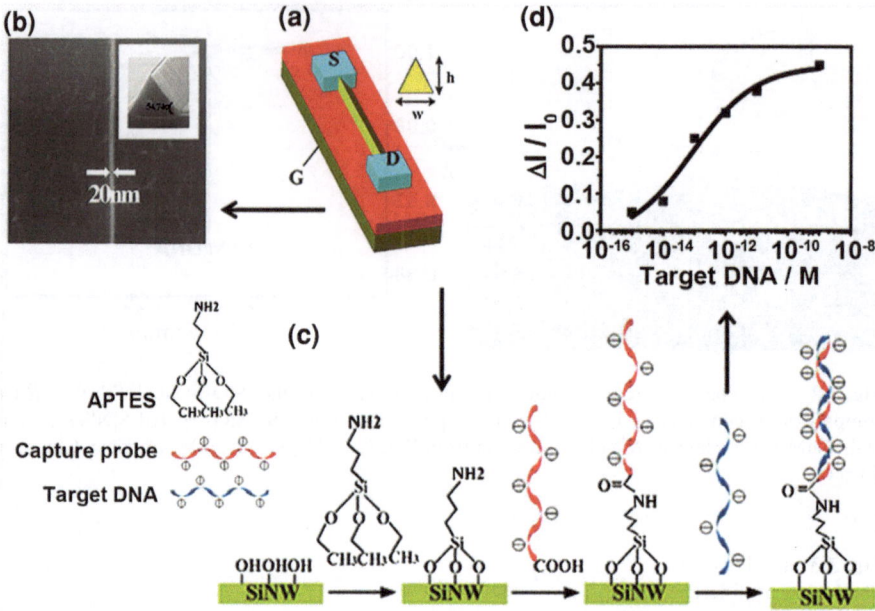

Fig. 3.4 a and **c** Schematics of SiNW-based sensor for DNA detection. **b** A typical SEM image of a SiNW (diameter <20 nm). The *inset* displays a cross-sectional view of the SiNW. **d** Linear relationship between the normalized current change ($|I / I_0|$) and target DNA concentration. I or I_0 stands for the value of the final or initial current, respectively. Reprinted with permission from Ref. [37]. Copyright 2011 American Chemical Society

Fig. 3.5 a Schematic diagram of RCA-based SiNW sensor for detecting DNA. **b** Relationship between the normalized current change and time with different target HBV DNA concentrations. Reprinted with permission from Ref. [41]. Copyright 2013 American Chemical Society

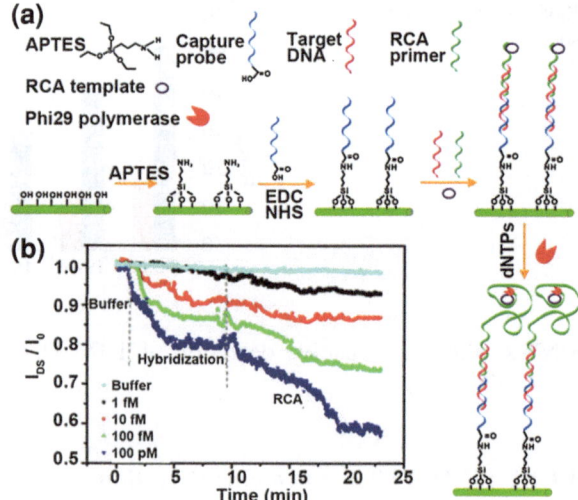

with a low detection limit of streptavidin (~ 10 pM) [15]. In their study, biotin modified on SiNWs surface was used as binding receptor. Addition of streptavidin led to a rapid increase of conductance of SiNWs, since the resultant biotin-modified SiNWs could specifically bind with streptavidin [68]. The conductance increased to a constant value at a threshold concentration of streptavidin [15]. To verify that the conductance was resulted from the specific binding between biotin and streptavidin, they further carried out several control experiments showing that few streptavidin molecules were nonspecifically linked with surface of bare SiNWs. This pioneer work provides valuable viewpoints for the design of high-performance SiNWs-FET devices.

Biomarker proteins are well recognized as specific indicators of diseases. Therefore, biomarkers detection can be utilized for disease screening, which is favorable for improving detection sensitivity of disease diagnosis. Considering that the blood merely contains ultralow concentration biomarkers, it is of essential importance to develop high-efficacy methods for rapid and precise detection of biomarkers in clinical diagnoses. In 2005, Lieber and coworkers realized multiplexing detection of various cancer markers by using the SiNW-FET sensor chip (Fig. 3.6a) [42]. The authors determined the sensitivity limits of the SiNW-FET devices by measuring conductance changes induced by various solution concentrations of prostate specific antigen (PSA), showing a linear relationship between conductance change and PSA concentrations ranging from ~ 5 ng/mL to 90 fg/mL (Fig. 3.6c). Moreover, this kind of SiNWs-based sensor featured sufficient multiplexed detection capabilities, since ~ 200 individual devices could be potentially included in the chip (Fig. 3.6b). As a result, three kinds of cancer biomarkers (e.g., PSA, carcinoma embryonic antigen (CEA), and mucin-1) were simultaneously detected with high sensitivity using the chip (Fig. 3.6d). Furthermore, the authors demonstrated the devices were also workable for specific detection in real systems (e.g., donkey and human serum samples). Following this work, other kinds of biomarkers were also detected using the SiNW-FET devices. For example, cardiac biomarker (human cardiac troponin-T, cTnT) of low concentrations down to $\sim$ fg/mL level was successfully detected in an assay buffer and an undiluted human serum environment [45].

Despite the above-mentioned success, there remains a major challenge for detection of the biomarkers for a whole blood sample using the SiNW-FET devices due to relatively severe non-specific absorption and a very short Debye length (λD) in the blood sample [61]. To address this issue, the physiological fluid samples should be prepurified before they flow into the SiNW-FET sensor. For example, Fahmy and coworkers introduced a microfluidic purification chip (MPC) system capable of preisolating targets, and then analyzed the treated samples (e.g., PSA and carbohydrate antigen 15.3 (CA 15.3)) using silicon "nanoribbons" FET devices (Fig. 3.7a) [69]. As schematically shown in Fig. 3.7 a, primary antibodies (e.g., anti-PSA and anti-CA15.3) modified on surface of the MPC were first specifically conjugated with biomarkers in the blood sample, followed by label-free detection using the SiNW-FET sensor, allowing sensitive detection of PSA of 2.5 ng/mL (Fig. 3.7b) and CA15.3 of 30 U/mL (Fig. 3.7c), respectively. Recently,

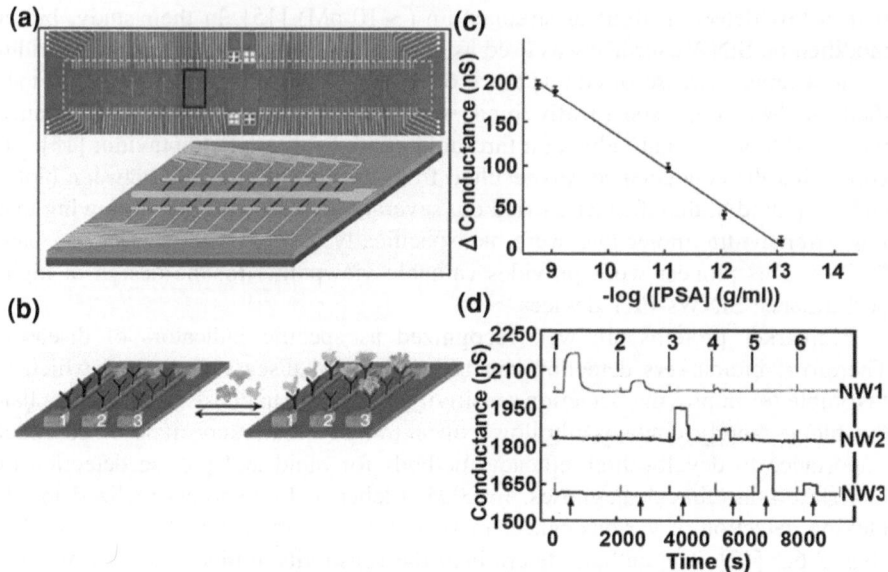

Fig. 3.6 **a** A typical picture (*top*) of the SiNW-based FET chip. The schematic (*bottom*) displays that SiNWs (*blue lines*) are connected with metal electrodes (*golden lines*). **b** Schematic representation of SiNWs-based sensing chip for multiplexed detection of three kinds of protein. **c** Linear relationship between the PSA concentrations and conductance response from the SiNWs-based sensor. **d** Simultaneous and multiplexed detection of three kinds of cancer biomarkers (e.g., PSA, CEA and mucin-1) according to observing the change of conductance in a real-time manner. The concentrations of the solution delivered to the chip are 0.9 ng/mL (1, PSA), 1.4 pg/mL (2, PSA), 0.2 ng/mL (3, CEA), 2 pg/mL (4, CEA), 0.5 ng/mL (5, mucin-1), and 5 pg/mL (6, mucin-1), respectively. Reprinted with permission from Nature Publishing Group, a division of Macmillan Publishers Ltd: Ref. [42], Copyright 2005

Patolsky and coworkers developed a kind of SiNWs-based separation and FET integrated chip, which was capable of rapid (~ 10 min) detection of the released target proteins from blood samples at a $\sim$pM level [70].

3.1.1.3 Other Applications

Besides detection of nucleic acids and proteins, other kinds of biological species and activity can also be readily detected using SiNW-FET devices, such as metal ions [15, 48], small molecules [50, 51], virus [52], cells [53], molecular interactions [54, 55], and cell's life activities [57–59]. Herein, we give a simple description of these applications as follows.

Lieber et al. developed a kind of SiNW-FET sensor arrays for selective detection of viruses at single-virus level by employing monoclonal antibody-modified p-type SiNWs arrays [52]. In this case, the discrete conductance changes were corresponding to the binding and unbinding of influenza A. The simultaneous

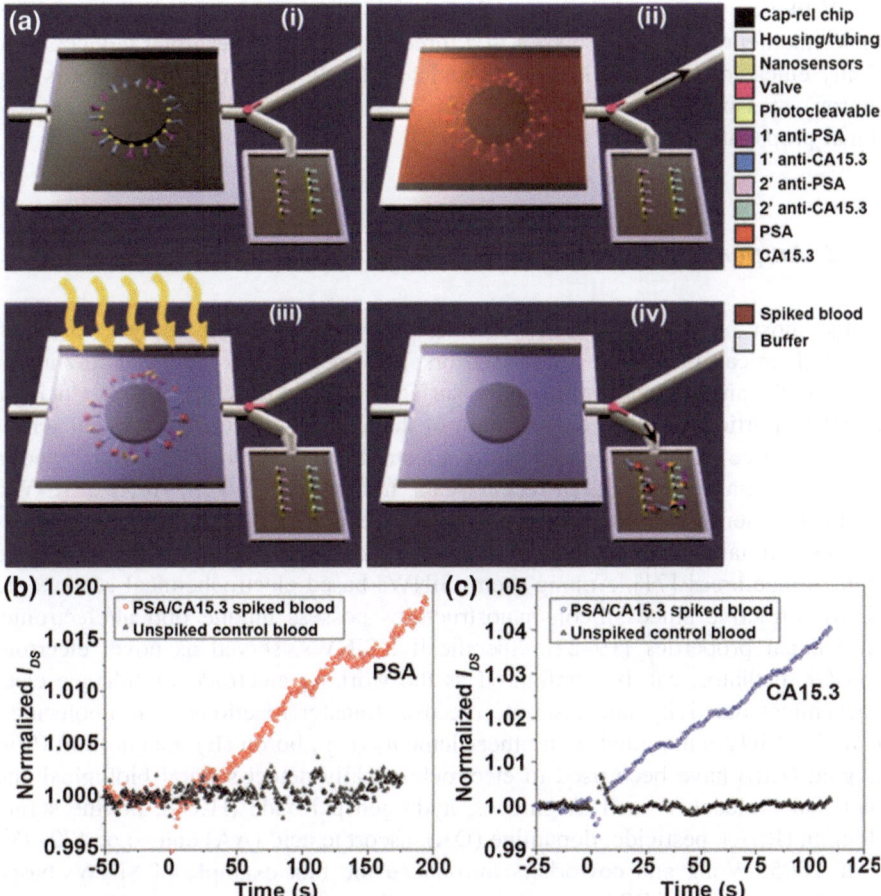

Fig. 3.7 a Schematic diagram of the MPC-FET for detection of biomarkers in the whole blood system. **b** and **c** Response of the FET modified with (**b**) anti-PSA- or (**c**) anti-CA15.3 for detection of PSA (2.5 ng/mL, (**b**)) or CA15.3 (30 U/mL, (**c**)) included in the blood system, respectively. Reprinted with permission from Nature Publishing Group, a division of Macmillan Publishers Ltd: Ref. [69], Copyright 2010

electric and optical measurements further confirmed that the conductance change was attributed to binding/unbinding of one single virus. Moreover, the prepared sensor devices modified with different antibody receptors possessed multiplexed detection capability, allowing multidetection of various viruses [52]. In 2008, Fahmy et al. fabricated a type of SiNW-FET sensors assisted by CMOS technology capable of rapid detection of antigen-specific T-cell [53]. Compared to other reported label-free approaches, this strategy showed much better sensitivity, with a detection limit of ~210 cells. Moreover, SiNW-FET arrays were also suitable for detection, stimulation, and inhibition of cellular bioelectricity, which was benefited for investigating signal propagation of neurons [57]. Typically, the

SiNW-FETs were patterned on a substrate and then passivated with polylysine, where the neurons could grow on in a direct manner. The resultant devices were highly efficacious for simultaneously and sensitively (50 'artificial synapses' per neuron) measuring abundant information (e.g., rate, amplitude, and shape) of signal propagation [57].

3.1.2 Amperometric-Based Biosensors

In the most basic terms, electrochemical sensors detect an analyte based on electrochemical reactions. These sensors are operated by either oxidizing or reducing the analyte in question on the surface of a working electrode, producing a signal proportional to the concentration of the analyte. Basically, a typical sensor contains three electrodes: a working electrode, a reference electrode, and a counter/auxiliary electrode. In recent years, nanomaterials (e.g., AuNPs, SiNWs, CNTs, graphene, etc.) have been extensively explored for construction of electrochemical nanosensors, taking advantage of their subtle electronic properties and high surface areas [71]. Among them, SiNWs-based electrochemical sensors are highly attractive since silicon nanostructures possess unique optical/electronic/mechanical properties [19–22]. Specifically, SiNWs, served as novel electron-transfer mediator, can be configured as the working electrode to enhance electrochemical reactivity and promote electron transfer reactions of biomolecules. Thus far, SiNWs modified with other elements (e.g., boron (B), magnesium (Mg) or gold (Au)) have been used to electrochemically detect several biological and chemical molecules, such as glucose, hydrogen peroxide (H_2O_2), bovine serum albumin (BSA), pesticide, dopamine (DA), ascorbic acid (AA) and so on [72–78].

In 2005, Wong and coworkers introduced the first example of SiNWs-based electrodes capable of BSA detection through the cyclic voltammetry (CV) method with high detection sensitivity (75 mA/mmol) [73]. In this case, SiNWs with large surface areas facilitated the immobilization of various analytes and the enhancement of electrical conductivity. Moreover, the SiNWs films modified with B or Mg were ready for the measurement of glucose or hydrogen peroxide, respectively, featuring a wide linear range (0–10 mM glucose) and extremely high sensitivity (172 nA/mmol) [72]. Very recently, He et al. employed AuNPs-coated SiNWs array as new nanoelectrode assembly for fabricating high-quality electrochemical sensors [78]. The resultant nanosensors enabled sensitive detection of different electroactive molecules (e.g., DA, AA, H_2O_2, and glucose). Typically, the detection limit for DA was as low as 40 nM, superior to that reported using Au electrode modified with AuNPs (220 nM) [79].

Another promising strategy is to employ SiNWs as the electron-transfer mediator and immobilization substrate to fabricate an amperometric enzyme-based biosensor, simultaneously featuring enzyme specificity and convenience of electro-analytical means [74–76]. For example, Lee and coworkers developed an electrochemical pesticide biosensor by immobilizing acetylcholinesterase (AChE)

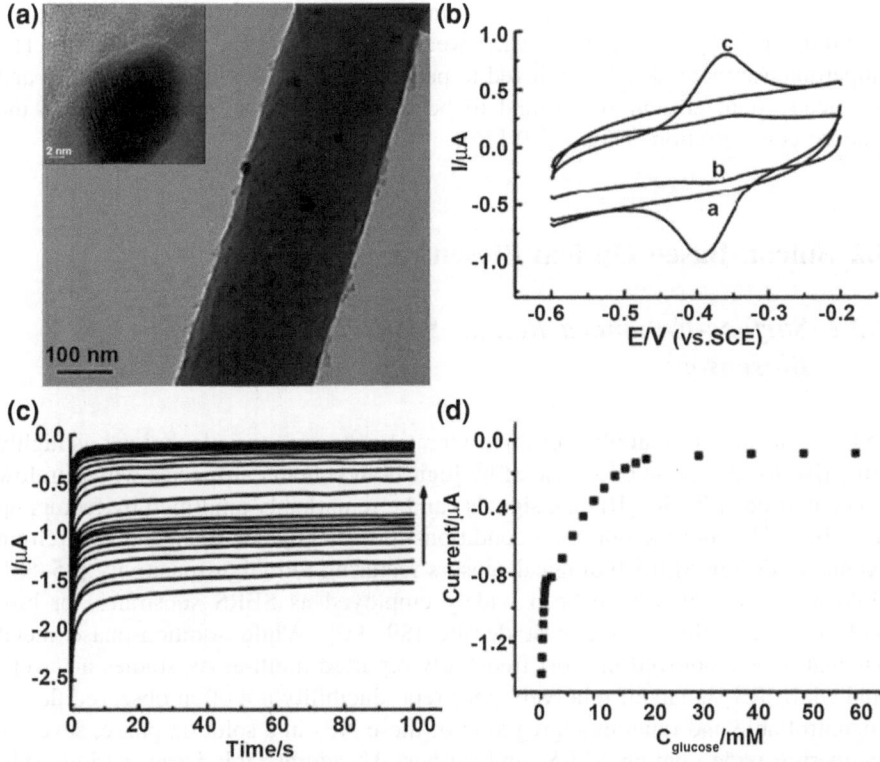

Fig. 3.8 **a** TEM and HRTEM (*inset*) images of the prepared one single AuNPs-decorated silicon nanowire. **b** Cyclic voltammograms of Nafion-AuNPs@SiNWs/GC (*curve a*), Nafion-GOx/GC (*curve b*) and Nafion-GOx-AuNPs@SiNWs/GC (*curve c*) electrodes in 0.1 M PBS (pH 6.0) saturated with N_2. *Scan rate* 50 mV/s. **c** Amperometric *i–t curves* of the Nafion-GOx-AuNPs@SiNWs/GC electrode for various glucose concentrations. **d** Calibration *curve* of the Nafion-GOx-AuNPs@SiNWs/GC electrode. Reproduced from Ref. [76] with permission from the Royal Society of Chemistry

on an electrode made of SiNWs coated with gold nanoparticles (AuNPs) (AuNPs@SiNWs) (Fig. 3.8a) [76]. Such AChE-based sensors were highly efficacious for rapidly detecting organophosphorous (OP) with enhanced enzymatic activity of AChE. As a result, the resultant sensing device exhibited high sensitivity, enabling detection of low-concentration (8 ng/L) dichlorvos (DDV), an OP pesticide [75]. In 2010, the same group further reported another electrochemical biosensor for glucose detection based on AuNPs@SiNWs with a low detection limit of 50 μM [76]. In this case, the immobilized GOx displayed two distinct redox peaks at − 0.360 and −0.393 V (Fig. 3.8b, curve c), which were ascribed to the facilitated redox reaction of flavin adenine dinucleotide (FAD) at the electrode surface. As controls, the Nafion-AuNPs@SiNWs/GC (Fig. 3.8b, curve a) did not show any electrochemical features in the cyclic voltammogram (CV), while Nafion-GOx/GC electrode only showed a pair of rather weak redox peaks (Fig. 3.8b, curve b). The amperometric method was employed to perform glucose detection (Fig. 3.8c) and the steady-state current was found to be reversely linearly proportional to the glucose concentrations (Fig. 3.8d) [76].

3.2 Silicon-Based Optical Biosensor

3.2.1 Surface-Enhanced Raman Scattering-Based Biosensors

SERS, considered as an ultrasensitive vibration spectroscopic technique, is highly attractive for detection of a trace of biological and chemical species with ultralow concentrations [80–84]. Raman signals can be remarkably amplified by factors up to $\sim 10^{12-14}$ under the optimum conditions in SERS, which opens up an exciting avenue to design SERS biological sensors featuring ultrahigh sensitivity [85–88]. Gold and silver NPs have been widely employed as SERS substrates for bio-molecules detection in the past decade [89, 90]. While solution-phase metal nanoparticles employed in most frequently reported multi-assay studies are well-studied, SERS signals of relatively poor reproducibility are often observed due to uncontrollable and random aggregation of these NPs in a solution phase, severely hampering wide-ranging SERS applications. To address this issue, various sub-strates (e.g., polymers, hydroxides, silicon nanomaterials, etc.) have been employed to prevent AgNPs/AuNPs aggregation by the formation of nanocom-posites. In particular, metallic nanoparticles-decorated silicon materials have been extensively studied. Notably, AgNPs/AuNPs aggregation can be efficiently pre-vented since AgNPs are steadily anchored on the surface of silicon materials [91–109]. For example, Lee et al. employed AgNPs/AuNPs-coated SiNWs as highly active SERS substrate (enhancement factor (EF): 10^{11}) for sensitive detection of a variety of chemical and biological species (e.g., Rhodamine 6G, crystal violet, nicotine and DNA) [97]. Besides, such AgNPs@SiNWs array was also suitable for ultrasensitive protein analysis and immunoassay, allowing detection of trace amounts (4 ng) of mouse immunoglobulinG and goat-anti-mouse immunoglobulin G [98]. Cheng and coworkers explored SiNWs arrays coated with silver as an efficient SERS substrate, allowing detection of Calcium dipicolinate with a low concentrations of $\sim 10-6$ M [91]. Boukherroub et al. used highly SERS-active (EF: $\sim 10^8$) AgNPs-capped SiNWs for sensitive detection of low-concentration $(10-1^4$ M) R6G [99]. Fang and coworkers deposited metal ions on the top of SiNWs to fabricate a SERS sensor, which was suitable for detecting ~ 600 mol-ecules [100]. Very recently, He et al. fabricated a sandwiched-structured SiNWs-based DNA SERS sensor, via serial immobilization of capture, target, and reported DNA, enabling ultrasensitive detection of DNA of a notably low concentration

(0.1 fM) [93]. They further developed a new kind of silicon-based SERS bio-sensors by employing AgNPs-decorated silicon wafers (AgNPs@Si) as SERS substrates [110]. Of particular significance, a large amount of AgNPs were tightly immobilized on Si surface, effectively avoiding random AgNPs aggregation. As a result, such silicon-based biosensor exhibited high reproducibility, allowing sensitive detection of DNA (~ 1 pM) with relatively small standard deviation (13.1 %).

The above-mentioned SERS methods generally involve "signal-on" procedure, i.e., signals are distinctly amplified once meeting the targets. Very recently, He and coworkers introduced a molecular beacon (MB)-assisted signal-off SERS strategies by using AuNPs-decorated SiNWs (AuNPs@SiNWAr) whose surface was modified with organic dyes-tagged stem-loop DNA [94]. As schematically illustrated in Fig. 3.9, the AuNPs@SiNWAr modified with the stem-loop oligonucleotide produces significant SERS intensities since the dye molecules are very close to AuNPs. Once hybridizing with target DNA, the stem-loop structure of the capture DNA changes to a linear duplex with strong rigidity. As a result, the distance between dye molecules and AuNPs become much longer, resulting in dramatic reduction of SERS intensities (so-called "signal-off" procedure). This novel signal-off SERS strategy was superbly suited to sensitive and multiplexed detection of DNA at a $\sim$pM level. More recently, the same authors further explored the AgNPs@Si as sensitive cellular sensors for sing-cell apoptosis detection in a noninvasive and label-free manner (Fig. 3.10) [111]. In their experiment, one single apoptotic cell was readily detected via SERS mapping techniques, suggesting novel possibilities for SERS sensing applications in vitro.

3.2.2 Fluorescence-Based Biosensors

In addition to SERS applications, SiNWs are also highly attractive for the design of fluorescence-based sensors. In 2008, Lee and coworkers designed a SiNWs fluorescent sensor for copper ions (Cu^{2+}) detection with high sensitivity and specificity [112]. In brief, the authors covalently modified SiNWs with N-(quin-oline-8-yl)-2-(3-triethoxysilyl-propylamino)-acetamide (QIOEt, served as fluorescence ligand). The increase of Cu^{2+} concentration led to the decrease of fluorescent intensities of the as prepared QIOEt-modified SiNWs, which was due to Cu^{2+}-induced fluorescence quenching of QIOEt. Based on the observed changes of fluorescence, low-concentration ($10-8$ M) Cu^{2+} could be readily detected. In addition to the high sensitivity, this kind of fluorescent sensor featured excellent specificity, allowing specific detection of Cu^{2+} in spite of existence of interfering metal ions [112]. More recently, He et al. employed AuNPs-coated SiNWs as novel quenchers with high fluorescence quenching efficiency (>90 %) to construct high-performance multicolor molecular beacons (MBs). Of particular significance, the resultant SiNWs-based MBs were highly stable in solutions with high salt concentrations (e.g., 0.1 M) and in a wide temperature range (e.g., 10–80 °C).

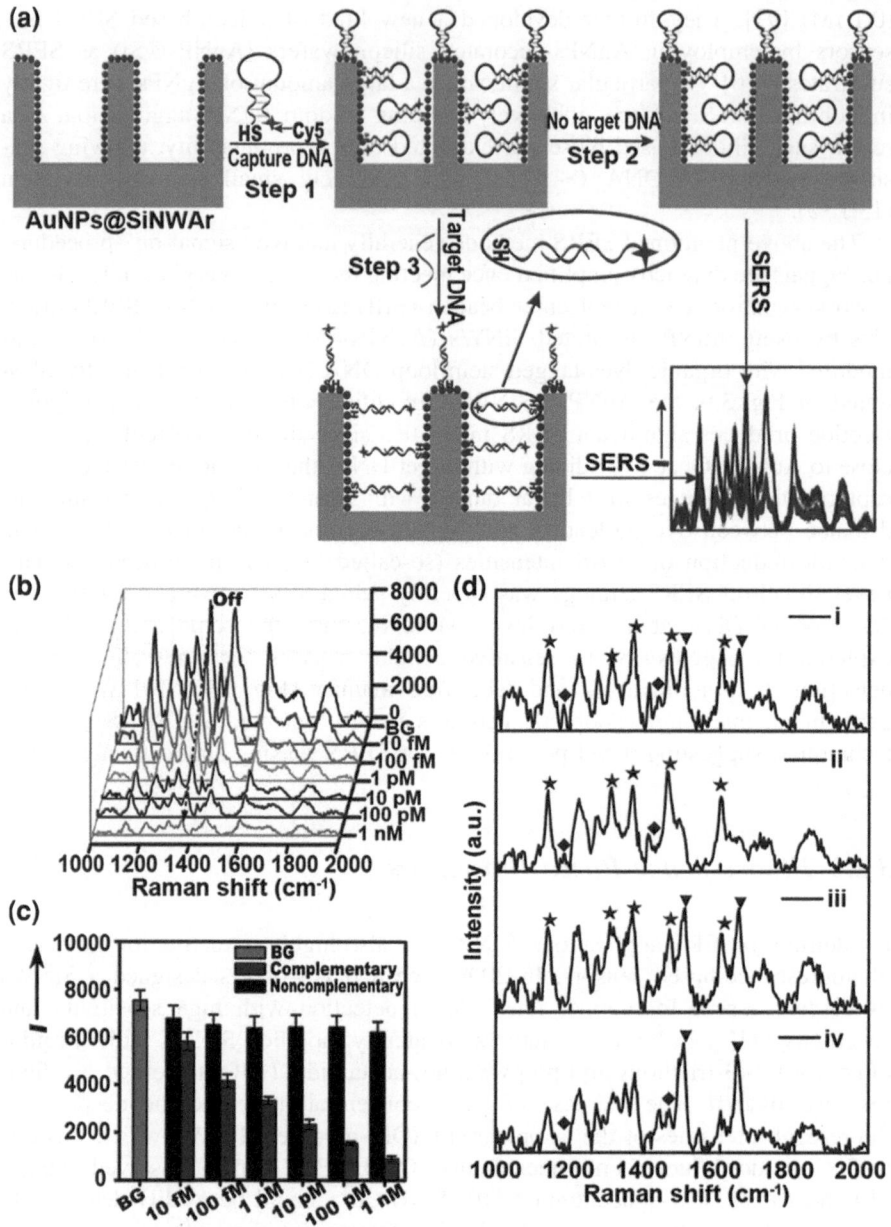

Fig. 3.9 a Schematic diagram of the signal-off SERS strategy. **b** SERS spectra and corresponding Raman intensity (**c**) of 1,366 cm^{-1} peak of target DNA with various concentrations. Background is shorted for *BG*. **d** Typical SERS spectra produced by the signal-off sensor for multiplexed detection of p53, p16, or p21. In particular (*i*) is the concurrent SERS spectra of p53, p16, and p21 coated the AuNPs@SiNWAr surface. (*ii–iv*) are individual SERS spectra of addition of p53, p16, or p21 target DNA with a concentration of 1 pM), respectively. Typical Raman peaks of ROX, FAM, and Cy5 are indicated *Triangle*, *quadrangle*, and *pentagon*, respectively. Reproduced from Ref. [94] with permission from John Wiley & Sons Inc

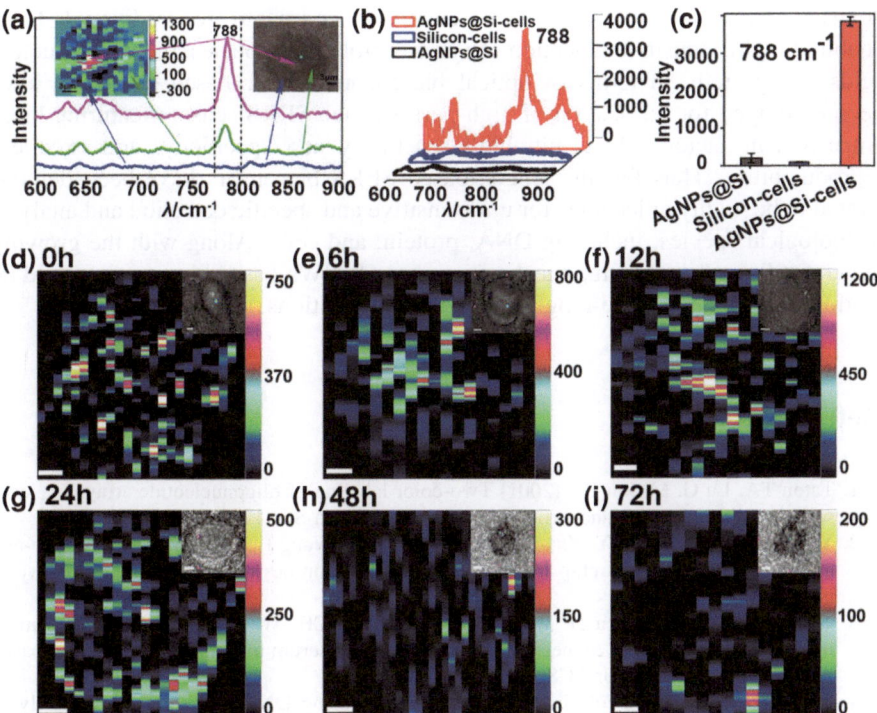

Fig. 3.10   a Typical SERS spectra of one single A549 cell cultured on surface of the AgNPs@Si substrate. Corresponding SERS mapping image and bright-filed image are shown in the *inset*. **b** and **c** show Raman spectra and corresponding SERS intensity of 788 cm^{-1} band of the pure AgNPs@Si, silicon wafer incubated with cells, and AgNPs@Si incubated with cells. **d–i** SERS mapping images of the A549 cells at different apoptotic stages. Distinct colors stand for various SERS intensity induced by DNA with different concentrations. Reprinted with permission from Ref. [111]. Copyright 2013 American Chemical Society

Besides, the silicon MBs were suitable for simultaneously assembling various DNA strands due to huge surface of the prepared AuNPs-coated SiNW, allowing sensitive and multiplexed DNA detection at ∼pM levels [113].

3.3   Conclusions and Prospects

In this chapter, we have summarized the representative works to highlight the recent developments of fabricating SiNWs-based biosensors for sensing applications. Thus far, great strides have been made in designing SiNWs-based biosensors in two categories, i.e., electrochemical biosensors and optical biosensors. In particular, SiNWs-based FET, as one of the most important electrochemical sensing devices, exhibits vastly improved sensitivity compared to traditional FET, enabling highly sensitive detection of various biological species, including nucleic acids and protein. In terms of optical biosensors, silicon-based substrates have been employed for fabrication of high-performance SERS sensors featuring large enhancement factor values, ultrahigh sensitivity and specificity, and excellent reproducibility. Thus far, the silicon-based SERS biosensors have been demonstrated to be highly efficacious for ultrasensitive and specific detection and analysis of biological species, including DNA, protein, and cells. Along with the growing understanding of the sensing mechanism, those SiNWs-based biosensors would be readily available for wide-ranging practical applications.

References

1. Taton TA, Lu G, Mirkin CA (2001) Two-color labeling of oligonucleotide arrays via size-selective scattering of nanoparticle probes. J Am Chem Soc 123(21):5164–5165
2. Yan J, Hu M, Li D, He Y, Zhao R, Jiang X, Song S, Wang L, Fan CH (2008) A nano-and micro-integrated protein chip based on quantum dot probes and a microfluidic network. Nano Res 1(6):490–496
3. Hu M, Yan J, He Y, Lu H, Weng L, Song S, Fan CH, Wang L (2009) Ultrasensitive, multiplexed detection of cancer biomarkers directly in serum by using a quantum dot-based microfluidic protein chip. ACS Nano 4(1):488–494
4. Liu J, Lu Y (2003) A colorimetric lead biosensor using DNAzyme-directed assembly of gold nanoparticles. J Am Chem Soc 125(22):6642–6643
5. Zhang J, Song S, Wang L, Pan D, Fan C (2007) A gold nanoparticle-based chronocoulometric DNA sensor for amplified detection of DNA. Nat Protoc 2(11):2888–2895
6. Song S, Qin Y, He Y, Huang Q, Fan C, Chen H-Y (2010) Functional nanoprobes for ultrasensitive detection of biomolecules. Chem Soc Rev 39(11):4234–4243
7. Giljohann DA, Mirkin CA (2009) Drivers of bio diagnostic development. Nature 462(7272):461–464
8. Chemla Y, Grossman H, Poon Y, McDermott R, Stevens R, Alper M, Clarke J (2000) Ultrasensitive magnetic biosensor for homogeneous immunoassay. Proc Natl Acad Sci USA 97(26):14268–14272
9. Perez JM, Josephson L, O'Loughlin T, Högemann D, Weissleder R (2002) Magnetic relaxation switches capable of sensing molecular interactions. Nat Biotechnol 20(8):816–820
10. Besteman K, Lee J-O, Wiertz FG, Heering HA, Dekker C (2003) Enzyme-coated carbon nanotubes as single-molecule biosensors. Nano Lett 3(6):727–730
11. Chen RJ, Bangsaruntip S, Drouvalakis KA, Kam NWS, Shim M, Li Y, Kim W, Utz PJ, Dai H (2003) Noncovalent functionalization of carbon nanotubes for highly specific electronic biosensors. Proc Natl Acad Sci USA 100(9):4984–4989
12. Wang J, Musameh M, Lin Y (2003) Solubilization of carbon nanotubes by Nafion toward the preparation of amperometric biosensors. J Am Chem Soc 125(9):2408–2409
13. Chen RJ, Choi HC, Bangsaruntip S, Yenilmez E, Tang X, Wang Q, Chang Y-L, Dai H (2004) An investigation of the mechanisms of electronic sensing of protein adsorption on carbon nanotube devices. J Am Chem Soc 126(5):1563–1568
14. Byon HR, Choi HC (2006) Network single-walled carbon nanotube-field effect transistors (SWNT-FETs) with increased Schottky contact area for highly sensitive biosensor applications. J Am Chem Soc 128(7):2188–2189

15. Cui Y, Wei Q, Park H, Lieber CM (2001) Nanowire nanosensors for highly sensitive and selective detection of biological and chemical species. Science 293(5533):1289–1292
16. Patolsky F, Lieber CM (2005) Nanowire nanosensors. Mater Today 8(4):20–28
17. Hu J, Odom TW, Lieber CM (1999) Chemistry and physics in one dimension: synthesis and properties of nanowires and nanotubes. Acc Chem Res 32(5):435–445
18. Duan X, Huang Y, Cui Y, Wang J, Lieber CM (2001) Indium phosphide nanowires as building blocks for nanoscale electronic and optoelectronic devices. Nature 409(6816):66–69
19. Kostarelos K, Bianco A, Prato M (2009) Promises, facts and challenges for carbon nanotubes in imaging and therapeutics. Nat Nanotechnol 4(10):627–633
20. Baughman RH, Zakhidov AA, de Heer WA (2002) Carbon nanotubes–the route toward applications. Science 297(5582):787–792
21. Ma D, Lee C, Au F, Tong S, Lee S-T (2003) Small-diameter silicon nanowire surfaces. Science 299(5614):1874–1877
22. Schmidt V, Wittemann JV, Senz S, Gösele U (2009) Silicon nanowires: a review on aspects of their growth and their electrical properties. Adv Mater 21(25–26):2681–2702
23. Thévenot DR, Toth K, Durst RA, Wilson GS (2001) Electrochemical biosensors: recommended definitions and classification. Biosensors Bioelectron 16(1):121–131
24. Grieshaber D, MacKenzie R, Vörös J, Reimhult E (2008) Electrochemical biosensors-sensor principles and architectures. Sensors 8(3):1400–1458
25. Chen K-I, Li B-R, Chen Y-T (2011) Silicon nanowire field-effect transistor-based biosensors for biomedical diagnosis and cellular recording investigation. Nano Today 6(2):131–154
26. Patolsky F, Zheng G, Lieber CM (2006) Nanowire sensors for medicine and the life sciences. Nanomedicine 1(1):51–65
27. Chen C-P, Ganguly A, Lu C-Y, Chen T-Y, Kuo C-C, Chen R-S, Tu W-H, Fischer WB, Chen K-H, Chen L-C (2011) Ultrasensitive in situ label-free DNA detection using a GaN nanowire-based extended-gate field-effect-transistor sensor. Anal Chem 83(6):1938–1943
28. Tian B, Xie P, Kempa TJ, Bell DC, Lieber CM (2009) Single-crystalline kinked semiconductor nanowire superstructures. Nat Nanotechnol 4(12):824–829
29. Cui Y, Duan X, Hu J, Lieber CM (2000) Doping and electrical transport in silicon nanowires. J Phys Chem B 104(22):5213–5216
30. Patolsky F, Zheng G, Lieber CM (2006) Fabrication of silicon nanowire devices for ultrasensitive, label-free, real-time detection of biological and chemical species. Nat Protoc 1(4):1711–1724
31. Hahm J-i, Lieber CM (2004) Direct ultrasensitive electrical detection of DNA and DNA sequence variations using nanowire nanosensors. Nano Lett 4(1):51–54
32. Li Z, Chen Y, Li X, Kamins TI, Nauka K, Williams RS (2004) Sequence-specific label-free DNA sensors based on silicon nanowires. Nano Lett 4(2):245–247
33. Bunimovich YL, Shin YS, Yeo W-S, Amori M, Kwong G, Heath JR (2006) Quantitative real-time measurements of DNA hybridization with alkylated nonoxidized silicon nanowires in electrolyte solution. J Am Chem Soc 128(50):16323–16331
34. Gao Z, Agarwal A, Trigg AD, Singh N, Fang C, Tung C-H, Fan Y, Buddharaju KD, Kong J (2007) Silicon nanowire arrays for label-free detection of DNA. Anal Chem 79(9):3291–3297
35. Cattani-Scholz A, Pedone D, Dubey M, Neppl S, Nickel B, Feulner P, Schwartz J, Abstreiter G, Tornow M (2008) Organophosphonate-based PNA-functionalization of silicon nanowires for label-free DNA detection. ACS Nano 2(8):1653–1660
36. Zhang G-J, Chua JH, Chee R-E, Agarwal A, Wong SM (2009) Label-free direct detection of MiRNAs with silicon nanowire biosensors. Biosens Bioelectron 24(8):2504–2508
37. Gao A, Lu N, Dai P, Li T, Pei H, Gao X, Gong Y, Wang Y, Fan CH (2011) Silicon-nanowire-based CMOS-compatible field-effect transistor nanosensors for ultrasensitive electrical detection of nucleic acids. Nano Lett 11(9):3974–3978

38. Gao A, Lu N, Wang Y, Dai P, Li T, Gao X, Wang Y, Fan CH (2012) Enhanced sensing of nucleic acids with silicon nanowire field effect transistor biosensors. Nano Lett 12(10):5262–5268
39. Dorvel BR, Reddy B, Go J, Duarte Guevara C, Salm E, Alam MA, Bashir R (2012) Silicon nanowires with high-k hafnium oxide dielectrics for sensitive detection of small nucleic acid oligomers. ACS Nano 6(7):6150–6164
40. Kwiat M, Elnathan R, Kwak M, de Vries JW, Pevzner A, Engel Y, Burstein L, Khatchtourints A, Lichtenstein A, Flaxer E, Herrmann A, Patolsky F (2012) Non-covalent monolayer-piercing anchoring of lipophilic nucleic acids: preparation, characterization, and sensing applications. J Am Chem Soc 134(1):280–292
41. Gao A, Zou N, Dai P, Lu N, Li T, Wang Y, Zhao J, Mao H (2013) Signal-to-noise ratio enhancement of silicon nanowires biosensor with rolling circle amplification. Nano Lett 13(9):4123–4130
42. Zheng G, Patolsky F, Cui Y, Wang WU, Lieber CM (2005) Multiplexed electrical detection of cancer markers with nanowire sensor arrays. Nat Biotechnol 23(10):1294–1301
43. Stern E, Klemic JF, Routenberg DA, Wyrembak PN, Turner-Evans DB, Hamilton AD, LaVan DA, Fahmy TM, Reed MA (2007) Label-free immunodetection with CMOS-compatible semiconducting nanowires. Nature 445(7127):519–522
44. Kim A, Ah CS, Yu HY, Yang J-H, Baek I-B, Ahn C-G, Park CW, Jun MS, Lee S (2007) Ultrasensitive, label-free, and real-time immunodetection using silicon field-effect transistors. Appl Phys Lett 91(10):103901–103903
45. Chua JH, Chee R-E, Agarwal A, Wong SM, Zhang G-J (2009) Label-free electrical detection of cardiac biomarker with complementary metal-oxide semiconductor-compatible silicon nanowire sensor arrays. Anal Chem 81(15):6266–6271
46. Gong J-R (2010) Label-free attomolar detection of proteins using integrated nanoelectronic and electrokinetic devices. Small 6(8):967–973
47. Huang Y-W, Wu C-S, Chuang C-K, Pang S-T, Pan T-M, Yang Y-S, Ko F-H (2013) Real-time and label-free detection of the prostate-specific antigen in human serum by a polycrystalline silicon nanowire field-effect transistor biosensor. Anal Chem 85(16):7912–7918
48. Luo L, Jie J, Zhang W, He Z, Wang J, Yuan G, Zhang W, Wu LCM, Lee S-T (2009) Silicon nanowire sensors for Hg^{2+} and Cd^{2+} ions. Appl Phys Lett 94(19):193101–193103
49. Lin C-H, Hsiao C-Y, Hung C-H, Lo Y-R, Lee C-C, Su C-J, Lin H-C, Ko F-H, Huang T-Y, Yang Y-S (2008) Ultrasensitive detection of dopamine using a polysilicon nanowire field-effect transistor. Chem Commun 44:5749–5751
50. McAlpine MC, Ahmad H, Wang D, Heath JR (2007) Highly ordered nanowire arrays on plastic substrates for ultrasensitive flexible chemical sensors. Nat Mater 6(5):379–384
51. Engel Y, Elnathan R, Pevzner A, Davidi G, Flaxer E, Patolsky F (2010) Supersensitive detection of explosives by silicon nanowire arrays. Angew Chem Int Ed Engl 49(38):6830–6835
52. Patolsky F, Zheng G, Hayden O, Lakadamyali M, Zhuang X, Lieber CM (2004) Electrical detection of single viruses. Proc Natl Acad Sci USA 101(39):14017–14022
53. Stern E, Steenblock ER, Reed MA, Fahmy TM (2008) Label-free electronic detection of the antigen-specific T-cell immune response. Nano Lett 8(10):3310–3314
54. Wang WU, Chen C, Lin K-h, Fang Y, Lieber CM (2005) Label-free detection of small-molecule-protein interactions by using nanowire nanosensors. Proc Natl Acad Sci USA 102(9):3208–3212
55. Lin S-P, Pan C-Y, Tseng K-C, Lin M-C, Chen C-D, Tsai C-C, Yu S-H, Sun Y-C, Lin T-W, Chen Y-T (2009) A reversible surface functionalized nanowire transistor to study protein–protein interactions. Nano Today 4(3):235–243
56. Zhang G-J, Huang MJ, Ang JAJ, Yao Q, Ning Y (2013) Label-free detection of carbohydrate-protein interactions using nanoscale field-effect transistor biosensors. Anal Chem 85(9):4392–4397

57. Patolsky F, Timko BP, Yu G, Fang Y, Greytak AB, Zheng G, Lieber CM (2006) Detection, stimulation, and inhibition of neuronal signals with high-density nanowire transistor arrays. Science 313(5790):1100–1104

58. Pui T-S, Agarwal A, Ye F, Balasubramanian N, Chen P (2009) CMOS-compatible nanowire sensor arrays for detection of cellular bioelectricity. Small 5(2):208–212

59. Timko BP, Cohen-Karni T, Yu G, Qing Q, Tian B, Lieber CM (2009) Electrical recording from hearts with flexible nanowire device arrays. Nano Lett 9(2):914–918

60. Elfström N, Juhasz R, Sychugov I, Engfeldt T, Karlström AE, Linnros J (2007) Surface charge sensitivity of silicon nanowires: size dependence. Nano Lett 7(9):2608–2612

61. Stern E, Wagner R, Sigworth FJ, Breaker R, Fahmy TM, Reed MA (2007) Importance of the Debye screening length on nanowire field effect transistor sensors. Nano Lett 7(11):3405–3409

62. Zhang G-J, Zhang G, Chua JH, Chee R-E, Wong EH, Agarwal A, Buddharaju KD, Singh N, Gao Z, Balasubramanian N (2008) DNA sensing by silicon nanowire: charge layer distance dependence. Nano Lett 8(4):1066–1070

63. Hutvágner G, McLachlan J, Pasquinelli AE, Bálint É, Tuschl T, Zamore PD (2001) A cellular function for the RNA-interference enzyme dicer in the maturation of the let-7 small temporal RNA. Science 293(5531):834–838

64. Lee Y, Ahn C, Han J, Choi H, Kim J, Yim J, Lee J, Provost P, Rådmark O, Kim S (2003) The nuclear RNase III Drosha initiates microRNA processing. Nature 425(6956):415–419

65. Lu J, Getz G, Miska EA, Alvarez-Saavedra E, Lamb J, Peck D, Sweet-Cordero A, Ebert BL, Mak RH, Ferrando AA (2005) MicroRNA expression profiles classify human cancers. Nature 435(7043):834–838

66. Schena M, Shalon D, Davis RW, Brown PO (1995) Quantitative monitoring of gene expression patterns with a complementary DNA microarray. Science 270(5235):467–470

67. Draghici S, Khatri P, Eklund AC, Szallasi Z (2006) Reliability and reproducibility issues in DNA microarray measurements. Trends Genet 22(2):101–109

68. Bayer EA, Wilchek M (1990) Biotin-binding proteins: overview and prospects. Methods Enzymol 184:49–51 (Academic Press)

69. Stern E, Vacic A, Rajan NK, Criscione JM, Park J, Ilic BR, Mooney DJ, Reed MA, Fahmy TM (2010) Label-free biomarker detection from whole blood. Nat Nanotechnol 5(2):138–142

70. Krivitsky V, Hsiung L-C, Lichtenstein A, Brudnik B, Kantaev R, Elnathan R, Pevzner A, Khatchtourints A, Patolsky F (2012) Si nanowires forest-based on-chip biomolecular filtering, separation and preconcentration devices: nanowires do it all. Nano Lett 12(9):4748–4756

71. Feigel IM, Vedala H, Star A (2011) Biosensors based on one-dimensional nanostructures. J Mater Chem 21(25):8940–8954

72. Shao MW, Shan YY, Wong NB, Lee ST (2005) Silicon nanowire sensors for bioanalytical applications: glucose and hydrogen peroxide detection. Adv Funct Mater 15(9):1478–1482

73. Shao M-W, Yao H, Zhang M-L, Wong N-B, Shan Y-Y, Lee S-T (2005) Fabrication and application of long strands of silicon nanowires as sensors for bovine serum albumin detection. Appl Phys Lett 87(18):183103–183106

74. Chen W, Yao H, Tzang CH, Zhu J, Yang M, Lee S-T (2006) Silicon nanowires for high-sensitivity glucose detection. Appl Phys Lett 88(21):213104

75. Su S, He Y, Zhang M, Yang K, Song S, Zhang X, Fan CH, Lee S-T (2008) High-sensitivity pesticide detection via silicon nanowires-supported acetylcholinesterase-based electrochemical sensors. Appl Phys Lett 93(2):023113

76. Su S, He Y, Song S, Li D, Wang L, Fan CH, Lee S-T (2010) A silicon nanowire-based electrochemical glucose biosensor with high electrocatalytic activity and sensitivity. Nanoscale 2(9):1704–1707

77. Yan S, He N, Song Y, Zhang Z, Qian J, Xiao Z (2010) A novel biosensor based on gold nanoparticles modified silicon nanowire arrays. J Electroanal Chem 641(1–2):136–140

78. Su S, Wei X, Guo Y, Zhong Y, Su YY, Huang Q, Fan CH, He Y (2013) A silicon nanowire-based electrochemical sensor with high sensitivity and electrocatalytic activity. Part Part Syst Charact 30(4):326–331
79. Raj CR, Okajima T, Ohsaka T (2003) Gold nanoparticle arrays for the voltammetric sensing of dopamine. J Electroanal Chem 543(2):127–133
80. Fleischmann M, Hendra PJ, McQuillan AJ (1974) Raman spectra of pyridine adsorbed at a silver electrode. Chem Phys Lett 26(2):163–166
81. Albrecht MG, Creighton JA (1977) Anomalously intense Raman spectra of pyridine at a silver electrode. J Am Chem Soc 99(15):5215–5217
82. Jeanmaire DL, Van Duyne RP (1977) Surface Raman spectroelectrochemistry: part I. Heterocyclic, aromatic, and aliphatic amines adsorbed on the anodized silver electrode. J Electroanal Chem 84(1):1–20
83. Nie S, Emory SR (1997) Probing single molecules and single nanoparticles by surface-enhanced Raman scattering. Science 275(5303):1102–1106
84. Kneipp K, Wang Y, Kneipp H, Perelman LT, Itzkan I, Dasari RR, Feld MS (1997) Single molecule detection using surface-enhanced Raman scattering (SERS). Phys Rev Lett 78(9):1667–1670
85. Cao YC, Jin R, Mirkin CA (2002) Nanoparticles with Raman spectroscopic fingerprints for DNA and RNA detection. Science 297(5586):1536–1540
86. Qian X, Peng X-H, Ansari DO, Yin-Goen Q, Chen GZ, Shin DM, Yang L, Young AN, Wang MD, Nie S (2008) In vivo tumor targeting and spectroscopic detection with surface-enhanced Raman nanoparticle tags. Nat Biotechnol 26(1):83–90
87. Graham D, Thompson DG, Smith WE, Faulds K (2008) Control of enhanced Raman scattering using a DNA-based assembly process of dye-coded nanoparticles. Nat Nanotechnol 3(9):548–551
88. Chen Z, Tabakman SM, Goodwin AP, Kattah MG, Daranciang D, Wang X, Zhang G, Li X, Liu Z, Utz PJ, Jiang K, Fan S, Dai H (2008) Protein microarrays with carbon nanotubes as multicolor Raman labels. Nat Biotechnol 26(11):1285–1292
89. Vendrell M, Maiti KK, Dhaliwal K, Chang Y-T (2013) Surface-enhanced Raman scattering in cancer detection and imaging. Trends Biotechnol 31(4):249–257
90. Wang Y, Yan B, Chen L (2013) SERS tags: novel optical nanoprobes for bioanalysis. Chem Rev 113(3):1391–1428
91. Zhang B, Wang H, Lu L, Ai K, Zhang G, Cheng X (2008) Large-area silver-coated silicon nanowire arrays for molecular sensing using surface-enhanced Raman spectroscopy. Adv Funct Mater 18(16):2348–2355
92. He Y, Fan CH, Lee S-T (2010) Silicon nanostructures for bioapplications. Nano Today 5(4):282–295
93. He Y, Su S, Xu T, Zhong Y, Zapien JA, Li J, Fan CH, Lee S-T (2011) Silicon nanowires-based highly-efficient SERS-active platform for ultrasensitive DNA detection. Nano Today 6(2):122–130
94. Wei X, Su S, Guo Y, Jiang X, Zhong Y, Su YY, Fan CH, Lee S-T, He Y (2013) A molecular beacon-based signal-off surface-enhanced Raman scattering strategy for highly sensitive, reproducible, and multiplexed DNA detection. Small 9(15):2493–2499
95. Leng W, Yasseri AA, Sharma S, Li Z, Woo HY, Vak D, Bazan GC, Kelley AM (2006) Silver nanocrystal-modified silicon nanowires as substrates for surface-enhanced Raman and hyper-Raman scattering. Anal Chem 78(17):6279–6282
96. Hakim MMA, Lombardini M, Sun K, Giustiniano F, Roach PL, Davies DE, Howarth PH, de Planque MRR, Morgan H, Ashburn P (2012) Thin film polycrystalline silicon nanowire biosensors. Nano Lett 12(4):1868–1872
97. Shao M-W, Zhang M-L, Wong N-B, Ma DD-d, Wang H, Chen W, Lee S-T (2008) Ag-modified silicon nanowires substrate for ultrasensitive surface-enhanced Raman spectroscopy. Appl Phys Lett 93(23):233113–233118

98. Zhang M-L, Yi C-Q, Fan X, Peng K-Q, Wong N-B, Yang M-S, Zhang R-Q, Lee S-T (2008) A surface-enhanced Raman spectroscopy substrate for highly sensitive label-free immunoassay. Appl Phys Lett 92(4):043113–043116

99. Galopin E, Barbillat J, Coffinier Y, Szunerits S, Patriarche G, Boukherroub R (2009) Silicon nanowires coated with silver nanostructures as ultrasensitive interfaces for surface-enhanced Raman spectroscopy. ACS Appl Mater Int 1(7):1396–1403

100. Fang C, Agarwal A, Widjaja E, Garland MV, Wong SM, Linn L, Khalid NM, Salim SM, Balasubramanian N (2009) Metallization of silicon nanowires and SERS response from a single metallized nanowire. Chem Mater 21(15):3542–3548

101. Zhang M-L, Fan X, Zhou H-W, Shao M-W, Zapien JA, Wong N-B, Lee S-T (2010) A high-efficiency surface-enhanced Raman scattering substrate based on silicon nanowires array decorated with silver nanoparticles. J Phys Chem C 114(5):1969–1975

102. Wang XT, Shi WS, She GW, Mu LX, Lee S-T (2010) High-performance surface-enhanced Raman scattering sensors based on Ag nanoparticles-coated Si nanowire arrays for quantitative detection of pesticides. Appl Phys Lett 96(5):053104

103. Yi C, Li C-W, Fu H, Zhang M, Qi S, Wong N-B, Lee S-T, Yang M (2010) Patterned growth of vertically aligned silicon nanowire arrays for label-free DNA detection using surface-enhanced Raman spectroscopy. Anal Bioanal Chem 397(7):3143–3150

104. Peng Z, Hu H, Utama MIB, Wong LM, Ghosh K, Chen R, Wang S, Shen Z, Xiong Q (2010) Heteroepitaxial decoration of Ag nanoparticles on Si nanowires: a case study on Raman scattering and mapping. Nano Lett 10(10):3940–3947

105. Han X, Wang H, Ou X, Zhang X (2012) Highly sensitive, reproducible, and stable SERS sensors based on well-controlled silver nanoparticle-decorated silicon nanowire building blocks. J Mater Chem 22(28):14127–14132

106. Chen R, Li D, Hu H, Zhao Y, Wang Y, Wong N, Wang S, Zhang Y, Hu J, Shen Z, Xiong Q (2012) Tailoring optical properties of silicon nanowires by Au nanostructure decorations: enhanced Raman scattering and photodetection. J Phys Chem C 116(7):4416–4422

107. Han X, Wang H, Ou X, Zhang X (2013) Silicon nanowire-based surface-enhanced Raman spectroscopy endoscope for intracellular pH detection. ACS Appl Mater Interfaces 5(12):5811–5814

108. Wang H, Han X, Ou X, Lee C-S, Zhang X, Lee S-T (2013) Silicon nanowire based single-molecule SERS sensor. Nanoscale 5(17):8172–8176

109. Yang X, Zhong H, Zhu Y, Shen J, Li C (2013) Ultrasensitive and recyclable SERS substrate based on Au-decorated Si nanowire arrays. Dalton Trans 42(39):14324–14330

110. Jiang Z, Jiang X, Su S, Wei X, Lee S-T, He Y (2012) Silicon-based reproducible and active surface-enhanced Raman scattering substrates for sensitive, specific, and multiplex DNA detection. Appl Phys Lett 100(20):203104

111. Jiang X, Jiang Z, Xu T, Su S, Zhong Y, Peng F, Su Y, He Y (2013) Surface-enhanced Raman scattering-based sensing in vitro: facile and label-free detection of apoptotic cells at the single-cell level. Anal Chem 85(5):2809–2816

112. Mu L, Shi W, Chang JC, Lee S-T (2008) Silicon nanowires-based fluorescence sensor for Cu (II). Nano Lett 8(1):104–109

113. Su S, Wei X, Zhong Y, Guo Y, Su Y, Huang Q, Lee S-T, Fan C, He Y (2012) Silicon nanowire-based molecular beacons for high-sensitivity and sequence-specific DNA multiplexed analysis. ACS Nano 6(3):2582–2590

Chapter 4
Silicon-Based Nanoprobes for Bioimaging Applications

Abstract Bioimaging, serving as one of the most important techniques, provides direct visualization of biological systems. Biological probes are essential tools for bioimaging applications. Scientists have devoted tremendous efforts to developing various kinds of fluorescent nanomaterials (e.g., II–VI semiconductor quantum dots, fluorescent carbon nanodots, fluorescent nanodiamonds, silicon nanoparticles (SiNPs))-based bioprobes, significantly facilitating the advancement of bioimaging applications. Among them, SiNPs are regarded as potentially ideal fluorescent bioprobes due to their unique optical properties (e.g., high fluorescence and strong antiphotobleaching property) and excellent biocompatibility. In the past decade, SiNPs-based fluorescent bioprobes have been extensively explored for in vitro and in vivo imaging. Of particular significance, taking advantage of their ultrahigh photostability and non or lowly toxic properties, fluorescent SiNPs-based nanoprobes are demonstrated to be superbly suited to long-term and real-time bioimaging applications. Also of note, multifunctional SiNPs with fluorescent and magnetic properties have been developed for multimodel bioimaging studies in recent years.

Keywords Bioimaging · Biological fluorescent probes · Silicon nanoparticles · In vitro and in vivo · Magnetic resonance imaging · Tumor targeting · Cancer diagnosis

Fluorescence optical imaging is known as a powerful noninvasive manner for biological and biomedical studies from living cells to animals. Fluorescent probes are basically essential for bioimaging applications, including labeling the targeted molecules, investigating in vitro and in vivo behaviors of chemical and biological species, as well as diagnosing disease, etc. The quality of fluorescence imaging depends significantly on the fluorescent probes used in this technique. For optimum imaging and tracking of biological molecules, high-quality fluorescent probes are expected to be highly fluorescent, water-dispersible, chemical-/photostable, and biocompatible [1]. Three scientists, i.e., Osamu Shimomura, Martin Chalfie, and Roger Y. Tsien, won the Nobel Prize in Chemistry in 2008 owing to their discovery of an important kind of fluorescent bioprobes, i.e., green fluorescent protein (GFP).

Y. He and Y. Su, *Silicon Nano-biotechnology*, SpringerBriefs in Molecular Science,   61
DOI: 10.1007/978-3-642-54668-6_4, © The Author(s) 2014

To date, fluorescent proteins and organic dyes, serving as the first generation of fluorescent bioprobes, have been employed for wide-ranging biological and bio-medical research. However, these conventional bioprobes generally suffer from severe photobleaching, seriously limiting their applications, particularly for long-term bioimaging [2]. Since the first report on fluorescent II–VI quantum dots (QDs, also called semiconductor nanocrystals)-based fluorescent nanoprobes, they have appeared as a novel kind of high-performance biological nanoprobes for wide-ranging biological applications due to their unique optical properties (e.g., high fluorescence, robust photostability, size-tunable emission wavelengths, and broad photoexcitation coupled with narrow emission spectra, etc.) [1–7]. To date, these QDs-based bioprobes have been widely employed for a variety of bioimaging applications, including cellular labeling, in vivo animal targeting, and bioanalytical assays, etc. It is worthwhile to point out that, most II–VI QDs contain heavy metal ions (e.g., Cd^{2+}, Te^{2-}, etc.), leading to potential hazards in vitro and in vivo [8–14]. While several strategies of surface modification (e.g., ZnS shell/silica/polymer coating) have been introduced to alleviate the toxicity, the safety concern is not completely resolved, remaining a major challenge thus far. Consequently, tremendous investigation has been performed to design high-quality fluorescent biological probes with more favorable biocompatibility during the past decade.

Silicon nanoparticles (SiNPs), as the most important silicon zero-dimensional structure featuring unique optical properties, have been widely utilized in electronics-related fields [15, 16]. Of particular note, porous silicon nanoparticles are recently revealed to be biodegradable and readily excluded from the body via renal clearance with undetectable toxicity in vivo [17, 18]; besides, silicon is known as a common trace element widely distributed in the earth and can be naturally found in numerous tissues, indicating favorable biocompatibility of the fluorescent SiNPs. Consequently, together with their high fluorescence and robust photostability, fluorescent SiNPs have been extensively developed as a novel kind of high-performance nanoprobes for bioimaging applications, which will be introduced in this chapter in a detailed way.

In Sect. 4.1, we present a detailed introduction of SiNPs-based fluorescent bioprobes for in vitro imaging applications, demonstrating that the SiNPs-based nanoprobes are highly efficacious for cellular imaging in a long-term and real-time manner to take advantage of ultrahigh photostability and strong fluorescence of SiNPs. In Sect. 4.2, we review representative recent progress on SiNPs-based bioprobes for in vivo imaging. In Sect. 4.3, we analyze major challenges facing this field today, and further talk about potential development direction in the future.

4.1 In Vitro Imaging

To bring SiNPs for bioimaging applications, surface modification should be first performed to render SiNPs water-dispersible. In 2004 and 2005, Ruckenstein's and Tillery's group independently reported two kinds of water-dispersed SiNPs-based fluorescent labels, and primarily employed them for cellular imaging [19, 20]. In 2009, He, Lee and coworkers introduced a class of hydrophilic polymer-coated SiNPs (also called silicon nanospheres, SiNSs)-based bioprobes featuring excellent aqueous dispersibility, strong fluorescence, and robust photostability [21, 22]. HEK293T human kidney cells labeled by the prepared SiNPs-based probes showed resolved red fluorescent signals of SiNPs (Fig. 4.1a). Furthermore, luminescent signals of the cells could be clearly observed under excitation at different wavelengths as the SiNPs had a broad absorption spectrum (Fig. 4.1b). Most significantly, this study presented the first demonstration of SiNPs probes-based long-term cell imaging. Typically, the SiNPs-tagged cells preserved stable and strong fluorescent signals during 20-min observation by laser-scanning confocal microscopy (LSCM). In marked contrast, the signals of cells labeled by fluorescein isothiocyanate (FITC, a kind of traditional fluorescence dye) rapidly quenched due to severe photobleaching properties of organic dyes (Fig. 4.1c). However, following studies [22, 23] revealed that these prepared SiNPs generally suffered from poor pH stability, that is, fluorescence of SiNPs was severely quenched as pH value changed. To address this issue, He et al. further developed thermal oxidation method to prepare oxidized SiNSs (O-SiNSs) featuring robust pH stability, whose fluorescence was stable in wide-ranging pH values extending from 2 to 12 [22]. The authors further conjugated O-SiNSs with antibody, and applied the prepared O-SiNWs/antibody conjugates for immunofluorescent cell imaging. Meanwhile, Swihart and coworkers introduced a kind of (MSiNPs) whose fluorescence was stable in the pH range of 2–10, and further utilized them for cellular imaging applications [23]. Later, Tilley and coworkers developed a multistepped chemical method to produce allylamine-capped water-dispersible SiNPs, which were further applied for HeLa cell imaging [24]. In that case, the bright blue fluorescence from the SiNPs was distributed uniformly in the cytoplasm. In contrast, the control cells without SiNPs treatment showed minimal fluorescence, suggesting that the fluorescence observed in the HeLa cells arose from the emission of SiNPs.

It is worthwhile to point out that, all the above-mentioned SiNPs are hydrophobic due to surface-covered hydrophobic ligands (e.g., Si-H bonds). As a result, relatively complicated post-treatment (e.g., hydrophilic polymer/organic molecules coating and micelle encapsulation) is required to improve aqueous dispersibility of the SiNPs. Moreover, chemical/physical properties are prone to be deteriorated by the surface modification. For example, SiNPs modified with arylic acid or allylamine possess meager pH stability, which is difficult to be conjugated with protein. SiNPs encapsulated with hydrophilic polymer or micelles are pH stable, nevertheless, they generally have large sizes (>50 nm), which are relatively difficult to be excluded from body (nanoparticles with hydrodynamic radius

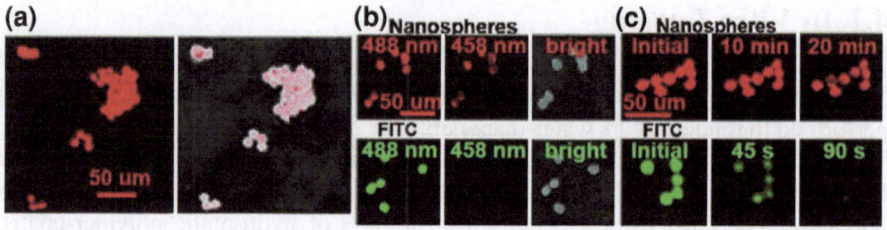

Fig. 4.1 **a** Confocal pictures of fluorescence SiNSs-labeled HEK-293T cells. **b** Fluorescent images of SiNWs- or FITC-tagged HEK-293T cells under different excitation wavelengths (e.g., 458 and 488 nm). **c** Confocal images of the SiNWs- or FITC-tagged HEK-293T cells during long-term continuous observation. "bright" stands for "bright field". Reproduced from Ref. [21] by permission of John Wiley & Sons Inc

smaller than 5 nm are revealed to be more easily cleared via renal clearance [25–28]). Besides, surface of SiNPs are probably destroyed by post-treatment (e.g., micelle encapsulation leads to dramatic decrease of quantum yield from ~17 to <10 %). Consequently, extensive efforts are in high demand to develop highly fluorescent, small-sized, stable and water-dispersible SiNPs for wide-ranging bioimaging applications. In 2011, He et al. employed SiNWs and glutaric acid as precursors to directly produce water-dispersed fluorescent SiNPs assisted by microwave irradiation [29]. Notably, the prepared SiNPs featured strong fluorescence, excellent aqueous dispersibility, small sizes (<5 nm in HD) and ultrahigh photostability. The authors further conjugated the as-prepared SiNPs with a goat anti-mouse antibody via an established EDC/NHS cross-linking reaction. The resultant SiNPs/antibody conjugates were then readily used for immunofluorescent targeting to microtubules of HeLa cells (one typical kind of cervical cancer cells) (Fig. 4.2a). Of particular note, taking advantage of ultrahigh photostability, the SiNPs were especially suitable for long-term cellular imaging. Typically, stable green fluorescence of SiNPs-targeted microtubules was observed during long-term (e.g., 120 min) continuous confocal observation (Fig. 4.2b). In marked contrast, fluorescent signals of the CdTe QDs-Fig. 4.2c) or FITC (Fig. 4.2d)-labeled cell were rapidly quenched during 25-min observation under the same conditions. In the following year, they further employed proteins as hydrophilic ligands for fabricating fluorescent and biofunctional SiNPs [30]. Remarkably, a large quantity of surface-covered hydrophilic protein molecules endowed the SiNPs with good water dispersibility and biospecific properties. Therefore, the as-prepared SiNPs could be directly utilized for immunofluorescence cell imaging without requiring additional complicated protein conjugation (Fig. 4.3).

Very recently, He et al. further introduced a novel facile bottom-up microwave method, enabling rapid large-scale preparation of highly fluorescent SiNPs using organosilicon molecules (e.g., 3-aminopropyl)trimethoxysilane, $C_6H_{17}NO_3Si$) as the silicon source [31]. Of particular significance, a large quantity of water-dispersible and fluorescent SiNPs could be rapidly produced in very short reaction time (e.g., 0.1 g SiNPs/10 min). Moreover, the resultant SiNPs featured bright

Fig. 4.2 a Photographs of immunofluorescent cell imaging using the SiNPs-based bioprobes. The SiNPs probes specifically target microtubule showing red signals. **b–d** Evolution fluorescence of SiNPs-(**b**), CdTe QDs-(**c**), and FITC (**d**)-labeled microtubules of one single HeLa cell. Reprinted with the permission from Ref. [29]. Copyright 2011 American Chemical Society

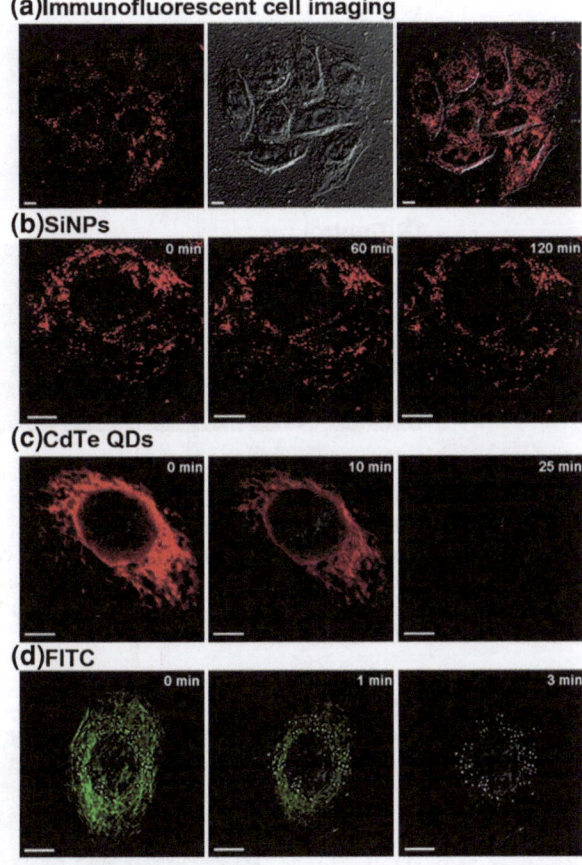

fluorescence (quantum yield: 20–25 %) and ultrahigh photostability, therefore were highly efficacious for immunofluorescent long-term cellular imaging. In their experiment, cell nuclei were stained with antibody-conjugated SiNPs, while cell microtubules were labeled with traditional dyes FITC. It could be found that, during a 60-min confocal observation, the photostable SiNPs-labeled nuclei yielded a steady fluorescent signal. In sharp contrast, for FITC-labeled microtubules, their green fluorescence quickly disappeared in only 3 min owing to poor anti-photobleaching property of organic dyes (Fig. 4.4). This work offers an invaluable avenue for design of high-performance SiNPs-based bioprobes for myriad bioimaging applications.

In addition to the above-mentioned exciting progress, SiNPs-based probes were recently also used to specifically target certain organelles, such as lysosomes, endoplasmic reticulum (ER), cytosol, and nuclei [32, 33]. For example, Ohta and coworkers demonstrated that lysosomes and ER could be selectively labeled by surface modified SiNPs [33]. In their work, the as-prepared SiNPs were treated

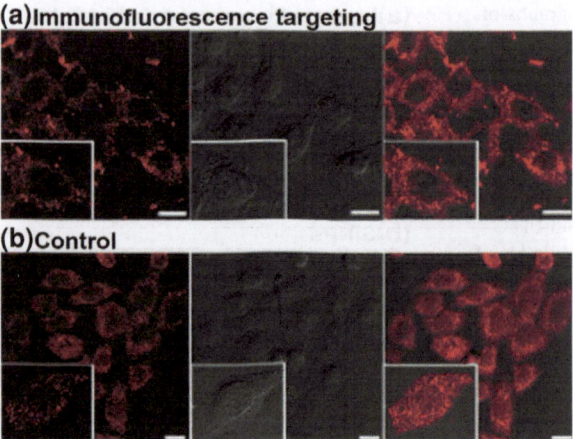

Fig. 4.3 Confocal pictures of HeLa cells imaged with the prepared SiNPs/protein bioconjugates (a) or pure SiNPs (b). In particular, the SiNPs/protein bioconjugates specifically target microtubules exhibiting *red* fluorescence. In marked contrast, for pure SiNPs-treat cells, the *red* fluorescent signals are uniformly distributed in the whole cellular region due to nonspecific absorption of the pure SiNPs. Scale bar stands for 10 μm. Insets present one single cell imaged using the prepared SiNPs/protein bioconjugates (a) and pure SiNPs (b). Reproduced from Ref. [30] by permission of John Wiley & Sons Inc

Fig. 4.4 a Confocal images of HeLa cells simultaneously labeled by SiNPs (nuclei, *left*) and FITC (microtubules, *middle*). **b** Temporal evolution of fluorescence signals of SiNPs (*blue*)- and FITC (*green*)-labeled cells during long-term continuous observation. Scale bars stand for 5 μm. Reprinted with the permission from Ref. [31]. Copyright 2013 American Chemical Society

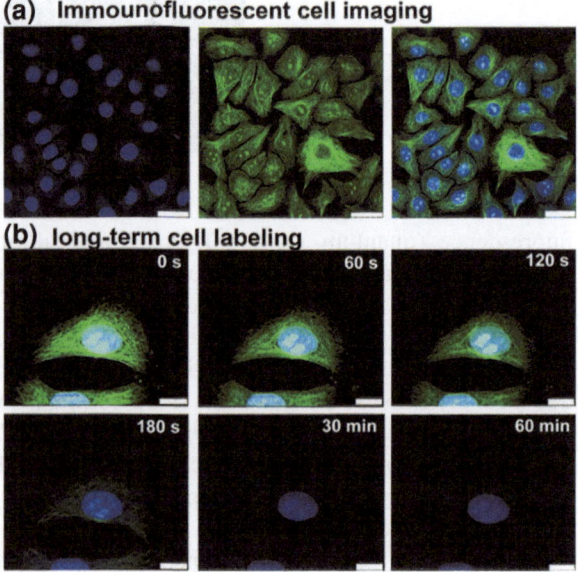

with two types of chemicals, i.e., allylamine and the amphiphilic block copolymer F127. It is worth noting that the modified SiNPs aggregates of different sizes (e.g., ~30 or 270 nm) could be obtained through controlling the amount of additives, allylamine, and F127. They found the prepared allylamine-modified SiNPs (named as SiNPs-Am) selectively labeled cell nuclei of the fixed human umbilical vein endothelial cells (HUVECs), while F127-modified SiNPs (named as SiNPs-F127) uniformly labeled cytosol. They further revealed that in the live HUVECs, SiNPs-Am selectively labeled the lysosomes no matter what the size was; while, in the case of SiNPs-F127, there was size-dependent intracellular localization, i.e., SiNPs-F127 with small size (named as SiNPs-F127$_{small}$) or larger size (named as SiNPs-F127$_{large}$) labeled the ER or lysosomes, respectively. The work indicates that some specific organelle imaging can be achieved by regulating the surface chemistry and sizes of SiNPs.

4.2 In Vivo Imaging

In addition to relatively sufficient reports concerning SiNPs-based bioprobes for cellular imaging in vitro, pioneer studies on in vivo imaging have been carried out in the past several years.

In 2011, Prasad and coworkers used micelle-encapsulated SiNPs (MSiNPs) for multiple cancer-related in vivo applications [34]. In this study, phospholipid micelles (e.g., 1,2-distearoyl-sn-glycero-3-phosphoethanolamine-N-[methoxy (polyethylene glycol)-2000) were used to modify SiNPs, producing the hydrophilic MSiNPs. In the in vivo tumor targeting work, the prepared MSiNPs were conjugated to RGD peptides (MSiNPs-RGD), and then intravenously injected into tumor-bearing nude mice. During 40-h postinjection, fluorescent intensities were increasingly enhanced at the tumor site, while feeble fluorescence was observed when the mice was injected with nonbioconjugated MSiNPs instead (Fig. 4.5a). As shown in Fig. 4.5b, MSiNPs-RGD injected tumor showed distinct red fluorescence of SiNPs, which was well consistent with the results shown in Fig. 4.5a. Based on quantitative analysis of fluorescent intensities, the MSiNPs-RGD treated tumor exhibited much stronger (186 times) fluorescence than that of the MSiNPs-treated group, indicating high effectiveness of tumor targeting using the MSiNPs-RGD. The authors also demonstrated the application of the MSiNPs for sentinel lymph node (SLN) mapping. Figure 4.5c shows the position of the SLN indicated by the prepared MSiNPs. Moreover, after subcutaneous injection, autofluorescence was well separated from the fluorescence of MSiNPs at various spots (Fig. 4.5d), which provided a possibility of using SiNPs as multicolor fluorescent bioprobes for multiplexed bioimaging applications. More recently, Jana et al. reported the low temperature colloid-chemical synthesis of 1–10 nm sized SiNPs exhibiting tunable visible emission wavelengths and ~6–13 % fluorescence quantum yield [35]. For bioimaging application, red-emitting hydrophobic SiNPs were modified by water-soluble folate molecules, and then used as fluorescent probes for specifically targeting cancer cells and tissues. The results showed that folate-functionalized SiNPs could selectively label cervical cancer positive tissue (highly over expressed

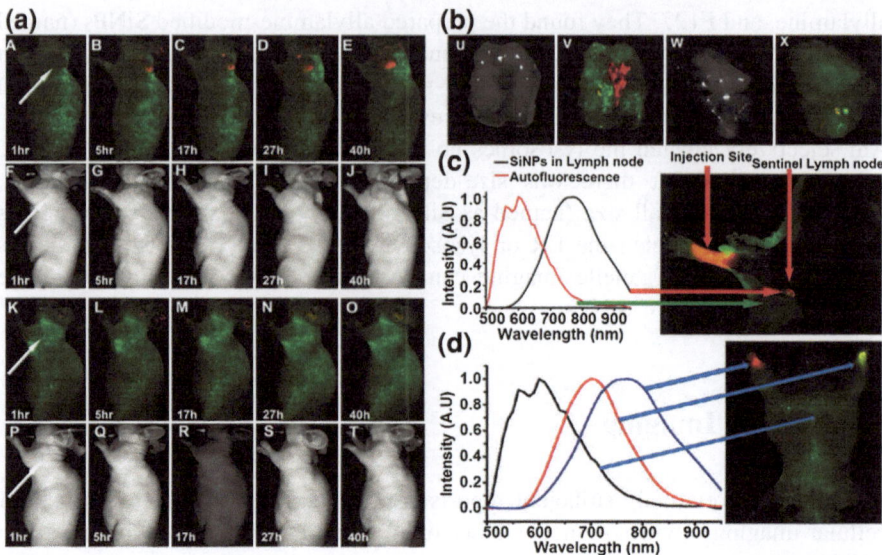

Fig. 4.5 a Real-time fluorescent images of Panc-1 tumor-bearing mice treated with (*A–E*) MSiNP-RGD or (*K–O*) MSiNPs. *Green* and *red* signals stand for autofluorescence of mice and the prepared SiNPs, respectively. *White images* in panels *A–E* and *K–O* are presented in Panels *F–J* and panels *P–T*, respectively. **b** Ex vivo pictures (*U*, *W*) and fluorescent pictures (*V*, *X*) of tumors tissues after 40-h p. j. of (*U*, *V*) MSiNP-RGD or (*W*, *X*) MSiNP. **c** Sentinel lymph node imaging in case of an axillary injected with the prepared MSiNPs. *Green* and *red* signals stand for autofluorescence of mice and the prepared SiNPs, respectively. **d** Fluorescence picture and corresponding PL spectra of a mouse subcutaneously injected with the SiNPs. *Green* and *red* signals stand for autofluorescence of mice and the prepared SiNPs, respectively. Reprinted with the permission from Ref. [34]. Copyright 2011 American Chemical Society

folate receptors); while the polymer-coated SiNPs without folate functionalization did not specifically label cancer positive tissue.

Magnetic resonance imaging (MRI) has been widely known as a noninvasive and high-resolution imaging technique for basic studies and clinical applications, especially suitable for in vivo imaging [36, 37]. Among them, silicon-relative nuclear magnetic resonance (NMR) has received much attention due to multihour nuclear spin relaxation (*T1*) times and dynamic nuclear polarization-induced hyperpolarization of bulk silicon [38–40]. Moreover, as the body naturally contains very little silicon, direct imaging of ^{29}Si is essentially background-free. Recently, a new technique has been demonstrated for positive-contrast MRI imaging using silicon-based micro and nanoparticles where the ^{29}Si nuclei within the silicon particles (4.7 %) are hyperpolarized and imaged directly in vivo using ^{29}Si MRI [41–44]. It is worth pointing out that, fluorescent and magnetic SiNPs doped with several kinds of magnetic materials (e.g., manganese (Mn), iron (Fe), or iron oxides (Fe_2O_3, Fe_3O_4)) [45–48] have been recently developed for MRI applications. In principle, the addition of paramagnetism to SiNPs would allow

Fig. 4.6 Fluorescent images and MRI of blank P388D1 cells (**a**), DS Si$_{Mn}$NPs-treated P388D1 cells (**b**) and D Si$_{Mn}$NPs-treated P388D1 macrophage cells (**c**). ([Mn$_{2+}$] $= 2 \times 10^{-5}$ M). Reprinted with the permission from Ref. [46]. Copyright 2010 American Chemical Society

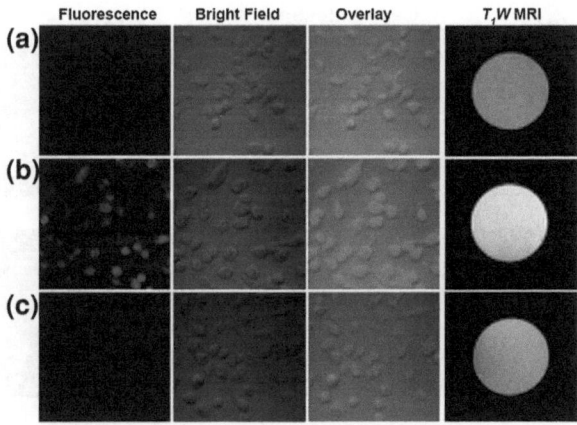

combinations of optical detection with MRI detection or magnetic separation. In addition to impurity doping with paramagnetic materials, an alternative means toward developing magnetic SiNPs is to co-encapsulate SiNPs with magnetic nanoparticles, such as Fe$_3$O$_4$ [49].

In 2009, Marcus and colleagues for the first time investigated (SiNPs) as a novel class of hyperpolarized MRI agent [41]. The authors revealed that silicon particles with different sizes ranging from 40 to 1 mm show distinctly long T_1 values (up to hours), much longer than those ever reported for hyperpolarized MRI imaging agents. In addition to the long T_1 value, the silicon particles were readily modified to endow them with required functionalities, thus serving as multifunctional imaging agents for MRI applications. In 2010, Tu and coworkers introduced water-soluble small-sized (core radius: ~ 2.15 nm) Mn-doped SiNPs (Si$_{Mn}$NPs) capable of concurrent MRI and two-photom dual-imaging [46]. As a result, the prepared Si$_{Mn}$NPs conjugated with dextran sulfate (DS) was able to specifically label scavenger receptors on macrophages, exhibiting strong fluorescence in the P388D1 murine macrophage cells (Fig. 4.6b), which is in sharp contrast to feeble fluorescence of the control group (i.e., cells without treatment of the probes) (Fig. 4.6a). Meanwhile, for MRI investigation, high contrast between DS-modified Si$_{Mn}$NPs (containing 0.02 mM Mn)-treated cell lysate and blank cell lysate was observed in $T1$-weighted (T_1W) MR images (Fig. 4.6a and b). In comparison, for the control group treated by pure Si$_{Mn}$NPs, both optical imaging and T_1W MRI images (Fig. 4.6c) showed that there was very limited uptake of DSi$_{Mn}$NPs compared to the targeted probes [46].

Very recently, Cassidy and coworkers utilized commercially purchased silicon particles (diameter: 2 μm) of high purity for direct in vivo imaging by MRI [42]. Notably, such hyperpolarized ^{29}Si MRI was versatile for a variety of in vivo imaging investigation. As shown in Fig. 4.7a, in case of gastrointestinal (GI) tract injected with the hyperpolarized silicon particles, detectable signals of ^{29}Si were observed. The particles were rapidly accumulated in the stomach and the

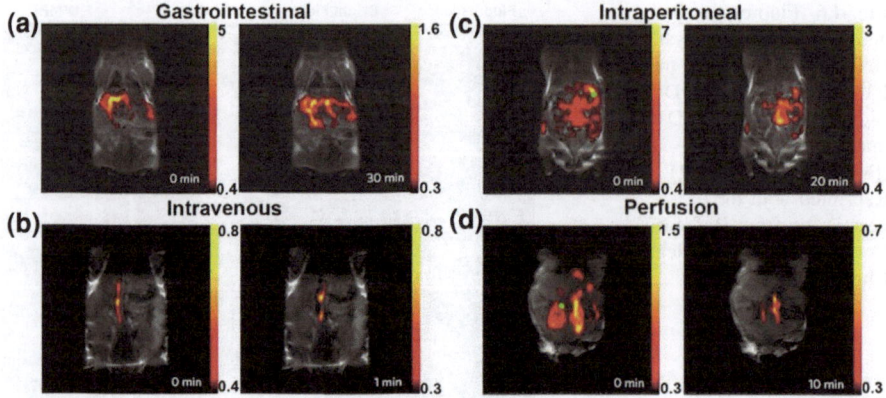

Fig. 4.7 Hyperpolarized silicon particles-assisted ^{29}Si MRI In vivo. **a–c** ^{29}Si MRI obtained through intragastric (**a**), intraperitoneal (**b**), intravenous, and (**c**), injection of the hyperpolarized silicon particles. **d** Perfusion imaging of the prostate tumor-bearing mouse administered with the hyperpolarized silicon particles. Reprinted by permission from Nature Publishing Group, a division of Macmillan Publishers Ltd: Ref. [42]. Copyright 2013

duodenum during 1-min postinjection (p.j.), and further reached the small intestines at 30-min p.j. Figure 4.7b shows clear structure of the GI tract once the silicon particles are delivered into the intraperitoneal (IP) cavity. The vena cava could be visibly and immediately imaged after intravenous injection of the particles, whose fluorescence was bright and stable after 1-min injection (Fig. 4.7c). In addition, the prostate tumor-bearing mouse was administered with the hyperpolarized silicon particles, exhibiting accurate mapping of the tumor, which provided a preliminary demonstration of the hyperpolarized ^{29}Si MRI-assisted perfusion imaging (Fig. 4.7d).

4.3   Conclusions and Prospects

In this chapter, recent representative progresses of silicon-based nanoprobes for bioimaging applications have been illustrated in a detailed way. In particular, a number of investigations demonstrated that fluorescent SiNPs as high-performance biological probes feature strong fluorescence, ultrahigh photostability, and favorable biocompatibility. As a result, these kinds of SiNPs-based bioprobes are highly efficacious for long-term and real-time in vitro and in vivo imaging studies, preserving bright and stable fluorescent signals for direct and long-time visualization of biological labeling. In addition to fluorescent bioimaging, multifunctional SiNPs simultaneously exhibiting fluorescent and magnetic properties have been developed for MRI with high spatial resolution and sensitivity. Despite exciting achievements on the utility of SiNPs for biological imaging, there still remain major challenges in the development of facile and low-cost strategies for designing high-performance SiNPs-based bioprobes, which is the precondition for wide-ranging biological and biomedical research. In addition, to improve bioimaging sensitivity, optical properties of the silicon-based probes should be further optimized via effective surface modification or novel fabrication techniques. Moreover, current studies mostly refer to single-color bioimaging; extensive efforts should be devoted to the fabrication of multiluminescent SiNPs-based bioprobes for multicolor in vitro and in vivo imaging.

References

1. Michalet X, Pinaud F, Bentolila L, Tsay J, Doose S, Li J, Sundaresan G, Wu A, Gambhir S, Weiss S (2005) Quantum dots for live cells, in vivo imaging, and diagnostics. Science 307(5709):538–544
2. Jaiswal JK, Mattoussi H, Mauro JM, Simon SM (2002) Long-term multiple color imaging of live cells using quantum dot bioconjugates. Nat Biotechnol 21(1):47–51
3. Bruchez M Jr, Moronne M, Gin P, Weiss S, Alivisatos AP (1998) Semiconductor nanocrystals as fluorescent biological labels. Science 281(5385):2013–2016
4. Chan WC, Nie S (1998) Quantum dot bioconjugates for ultrasensitive nonisotopic detection. Science 281(5385):2016–2018
5. Gao X, Cui Y, Levenson RM, Chung LWK, Nie S (2004) In vivo cancer targeting and imaging with semiconductor quantum dots. Nat Biotechnol 22(8):969–976
6. Medintz IL, Uyeda HT, Goldman ER, Mattoussi H (2005) Quantum dot bioconjugates for imaging, labelling and sensing. Nat Mater 4(6):435–446
7. Weissleder R, Kelly K, Sun EY, Shtatland T, Josephson L (2005) Cell-specific targeting of nanoparticles by multivalent attachment of small molecules. Nat Biotechnol 23(11):1418–1423
8. Derfus AM, Chan WCW, Bhatia SN (2004) Probing the cytotoxicity of semiconductor quantum dots. Nano Lett 4:11–18
9. Kirchner C, Liedl T, Kudera S, Pellegrino T, Munoz Javier A, Gaub HE, Stolzle S, Fertig N, Parak WJ (2005) Cytotoxicity of colloidal CdSe and CdSe/ZnS nanoparticles. Nano Lett 5(2):331–338
10. Jan E, Byrne SJ, Cuddihy M, Davies AM, Volkov Y, Gun'ko YK, Kotov NA (2008) High-content screening as a universal tool for fingerprinting of cytotoxicity of nanoparticles. ACS Nano 2(5):928–938
11. Su YY, He Y, Lu HT, Sai LM, Li QN, Li WX, Wang LH, Shen PP, Huang Q, Fan CH (2009) The cytotoxicity of cadmium based, aqueous phase-synthesized, quantum dots and its modulation by surface coating. Biomaterials 30(1):19–25
12. Hauck TS, Anderson RE, Fischer HC, Newbigging S, Chan WCW (2010) In vivo quantum-dot toxicity assessment. Small 6(1):138–144
13. Su Y, Hu M, Fan C, He Y, Li Q, Li W, Wang L-H, Shen P, Huang Q (2010) The cytotoxicity of CdTe quantum dots and the relative contributions from released cadmium ions and nanoparticle properties. Biomaterials 31:4829–4834
14. Chen N, He Y, Su YY, Li XM, Huang Q, Wang HF, Zhang XZ, Tai RZ, Fan CH (2012) The cytotoxicity of cadmium-based quantum dots. Biomaterials 33:1238–1244
15. He Y, Fan C, Lee S-T (2010) Silicon nanostructures for bioapplications. Nano Today 5(4):282–295

16. Kim B-H, Cho C-H, Park S-J, Park N-M, Sung GY (2006) Ni/Au contact to silicon quantum dot light-emitting diodes for the enhancement of carrier injection and light extraction efficiency. Appl Phys Lett 89:063509

17. Park J-H, Gu L, von Maltzahn G, Ruoslahti E, Bhatia SN, Sailor MJ (2009) Biodegradable luminescent porous silicon nanoparticles for in vivo applications. Nat Mater 8(4):331–336

18. Godin B, Gu J, Serda RE, Bhavane R, Tasciotti E, Chiappini C, Liu X, Tanaka T, Decuzzi P, Ferrari M (2010) Tailoring the degradation kinetics of mesoporous silicon structures through PEGylation. J Biomed Mater Res, Part A 94A(4):1236–1243

19. Li Z, Ruckenstein E (2004) Water-soluble poly (acrylic acid) grafted luminescent silicon nanoparticles and their use as fluorescent biological staining labels. Nano Lett 4(8):1463–1467

20. Warner JH, Hoshino A, Yamamoto K, Tilley RD (2005) Water-soluble photoluminescent silicon quantum dots. Angew Chem Int Ed 44(29):4550–4554

21. He Y, Kang ZH, Li QS, Tsang CHA, Fan CH, Lee S-T (2009) Ultrastable, highly fluorescent, and water-dispersed silicon-based nanospheres as cellular probes. Angew Chem Int Ed 48:128–132

22. He Y, Su Y, Yang X, Kang Z, Xu T, Zhang R, Fan C, Lee S-T (2009) Photo and pH stable, highly-luminescent silicon nanospheres and their bioconjugates for immunofluorescent cell imaging. J Am Chem Soc 131(12):4434–4438

23. Erogbogbo F, Yong K-T, Roy I, Xu G, Prasad PN, Swihart MT (2008) Biocompatible luminescent silicon quantum dots for imaging of cancer cells. ACS Nano 2(5):873–878

24. Shiohara A, Hanada S, Prabakar S, Fujioka K, Lim TH, Yamamoto K, Northcote PT, Tilley RD (2010) Chemical reactions on surface molecules attached to silicon quantum dots. J Am Chem Soc 132(1):248–253

25. Choi HS, Liu W, Misra P, Tanaka E, Zimmer JP, Ipe BI, Bawendi MG, Frangioni JV (2007) Renal clearance of quantum dots. Nat Biotechnol 25(10):1165–1170

26. Choi HS, Liu W, Liu F, Nasr K, Misra P, Bawendi MG, Frangioni JV (2009) Design considerations for tumour-targeted nanoparticles. Nat Nanotechnol 5(1):42–47

27. Su Y, Peng F, Jiang Z, Zhong Y, Lu Y, Jiang X, Huang Q, Fan C, Lee S-T, He Y (2011) In vivo distribution, pharmacokinetics, and toxicity of aqueous synthesized cadmium-containing quantum dots. Biomaterials 32(25):5855–5862

28. Lu Y, Su Y, Zhou Y, Wang J, Peng F, Zhong Y, Huang Q, Fan C, He Y (2013) In vivo behavior of near infrared-emitting quantum dots. Biomaterials 34(17):4302–4308. doi:10.1016/j.biomaterials.2013.02.054

29. He Y, Zhong Y, Peng F, Wei X, Su Y, Lu Y, Su S, Gu W, Liao L, Lee S-T (2011) One-pot microwave synthesis of water-dispersible, ultraphoto- and pH-stable, and highly fluorescent silicon quantum dots. J Am Chem Soc 133(36):14192–14195

30. Zhong Y, Peng F, Wei X, Zhou Y, Wang J, Jiang X, Su Y, Su S, Lee S-T, He Y (2012) Microwave-assisted synthesis of biofunctional and fluorescent silicon nanoparticles using proteins as hydrophilic ligands. Angew Chem Int Ed 51(34):8485–8489

31. Zhong Y, Peng F, Bao F, Wang S, Ji X, Yang L, Su Y, Lee S-T, He Y (2013) Large-scale aqueous synthesis of fluorescent and biocompatible silicon nanoparticles and their use as highly photostable biological probes. J Am Chem Soc 135(22):8350–8356

32. Shen P, Ohta S, Inasawa S, Yamaguchi Y (2011) Selective labeling of the endoplasmic reticulum in live cells with silicon quantum dots. Chem Commun 47(29):8409–8411

33. Ohta S, Shen P, Inasawa S, Yamaguchi Y (2012) Size-and surface chemistry-dependent intracellular localization of luminescent silicon quantum dot aggregates. J Mater Chem 22(21):10631–10638

34. Erogbogbo F, Yong K-T, Roy I, Hu R, Law W-C, Zhao W, Ding H, Wu F, Kumar R, Swihart MT, Prasad PN (2011) In vivo targeted cancer imaging, sentinel lymph node mapping and multi-channel imaging with biocompatible silicon nanocrystals. ACS Nano 5(1):413–423

35. Das P, Saha A, Maity AR, Ray SC, Jana NR (2013) Silicon nanoparticle based fluorescent biological label via low temperature thermal degradation of chloroalkylsilane. Nanoscale 5:5732–5737

36. Sipkins DA, Cheresh DA, Kazemi MR, Nevin LM, Bednarski MD, Li KCP (1998) Detection of tumor angiogenesis in vivo by $\alpha_v\beta_3$-targeted magnetic resonance imaging. Nat Med 4(5):623–626

37. Fox MD, Raichle ME (2007) Spontaneous fluctuations in brain activity observed with functional magnetic resonance imaging. Nat Rev Neurosci 8(9):700–711

38. Shulman RG, Wyluda BJ (1956) Nuclear magnetic resonance of ^{29}Si in n- and p-type silicon. Phys Rev 103(4):1127–1129

39. Ladd TD, Maryenko D, Yamamoto Y, Abe E, Itoh KM (2005) Coherence time of decoupled nuclear spins in silicon. Phys Rev B 71(1):014401

40. Dementyev AE, Cory DG, Ramanathan C (2008) Dynamic nuclear polarization in silicon microparticles. Phys Rev Lett 100(12):127601

41. Aptekar JW, Cassidy MC, Johnson AC, Barton RA, Lee M, Ogier AC, Vo C, Anahtar MN, Ren Y, Bhatia SN, Ramanathan C, Cory DG, Hill AL, Mair RW, Rosen MS, Walsworth RL, Marcus CM (2009) Silicon nanoparticles as hyperpolarized magnetic resonance imaging agents. ACS Nano 3(12):4003–4008

42. Cassidy MC, Chan HR, Ross BD, Bhattacharya PK, Marcus CM (2013) In vivo magnetic resonance imaging of hyperpolarized silicon particles. Nat Nanotechnol 8(5):363–368

43. Atkins TM, Cassidy MC, Lee M, Ganguly S, Marcus CM, Kauzlarich SM (2013) Synthesis of long T1 silicon nanoparticles for hyperpolarized ^{29}Si magnetic resonance imaging. ACS Nano 7(2):1609–1617

44. Ackerman JJ (2013) Magnetic resonance imaging: Silicon for the future. Nat Nanotechnol 8(5):313–315

45. Zhang X, Brynda M, Britt RD, Carroll EC, Larsen DS, Louie AY, Kauzlarich SM (2007) Synthesis and characterization of manganese-doped silicon nanoparticles: bifunctional paramagnetic-optical nanomaterial. J Am Chem Soc 129(35):10668–10669

46. Tu C, Ma X, Pantazis P, Kauzlarich SM, Louie AY (2010) Paramagnetic, silicon quantum dots for magnetic resonance and two-photon imaging of macrophages. J Am Chem Soc 132(6):2016–2023

47. Sato K, Yokosuka S, Takigami Y, Hirakuri K, Fujioka K, Manome Y, Sukegawa H, Iwai H, Fukata N (2011) Size-tunable silicon/iron oxide hybrid nanoparticles with fluorescence, superparamagnetism, and biocompatibility. J Am Chem Soc 133(46):18626–18633

48. Singh MP, Atkins TM, Muthuswamy E, Kamali S, Tu C, Louie AY, Kauzlarich SM (2012) Development of iron-doped silicon nanoparticles as bimodal imaging agents. ACS Nano 6(6):5596–5604

49. Erogbogbo F, Yong K-T, Hu R, Law W-C, Ding H, Chang C-W, Prasad PN, Swihart MT (2010) Biocompatible magnetofluorescent probes: luminescent silicon quantum dots coupled with superparamagnetic iron (III) oxide. ACS Nano 4(9):5131–5138

Chapter 5
Silicon-Based Nanoagents for Cancer Therapy

Abstract Nanotechnology has emerged as highly promising tools for cancer therapy. In comparison to traditional therapeutic methods (e.g., chemotherapy and radiotherapy) generally involving limited specificity and undesired side effects, nanomaterials (e.g., magnetic nanoparticles, carbon-based nanomaterials, and porous silicon, etc.)-based nanoagents and therapeutic strategies significantly facilitate the improvement of therapeutic efficacy and reduce the toxic side effect. In particular, silicon nanomaterials featuring large surface-to-volume ratio and abundant surface chemistry open up completely new avenues for design of novel silicon-based nanoagents with large drug-loading capacity and high tumor-targeting efficacy. In the past decade, porous silicon (pSi) has been widely employed as high-performance delivery systems for different types of therapeutic drug molecules, proteins, and genes, aiming at improved therapeutic efficacy and reduced toxic side effect. The pSi can be also functionalized with photosensitizers, radiosensitizers, or heat producers, which are efficacious for phototherapy or radiotherapy of cancer. On the other hand, silicon nanowires (SiNWs) and SiNWs-based nanohybrids (e.g., gold nanoparticles/nanospheres-coated SiNWs), serving as novel classes of drug nanocarriers and hyperthermia nanoagents, have recently been explored for in vitro and in vivo cancer therapy with encouraging therapeutic outcomes.

Keywords Cancer therapy · Chemotherapy · Phototherapy · Drug nanocarriers · Hyperthemia agents · Porous silicon · Silicon nanowires

Cancer is widely regarded as one of the most devastating malignant diseases, greatly threatening human health for many decades. As reported by the International Agency for Research on Cancer (IARC), the number of new cases of cancer is larger than 10 million each year, and will probably reach 21.4 million globally by 2030. Moreover, cancer leads to more than 6 million annual deaths to date [1, 2]. In past 50 years, human being has made vast efforts in cancer therapy [2]. Thus far, chemotherapy and radiotherapy are major therapeutic approaches for cancer treatment; they are nevertheless nonspecific to cancer cells and generally produce adverse effects on normal tissues and organs [3]. In particular, for cancer

Y. He and Y. Su, *Silicon Nano-biotechnology*, SpringerBriefs in Molecular Science,   75
DOI: 10.1007/978-3-642-54668-6_5, © The Author(s) 2014

chemotherapy, drug molecules have to pass through a number of transport barriers to reach the tumor site. Moreover, drugs are prone to be quickly cleared from body via kidney filtration and renal clearance, which is adverse to therapeutic efficiency [4]. In recent years, taking advantage of the huge surface-to-volume ratio and surface tailorability, a variety of nanomaterials (e.g., magnetic nanoparticles, carbon-based nanomaterials, and porous silicon, etc.) have been widely explored as novel drugs or vaccine carriers [3].

Silicon nanomaterials featuring large surface-to-volume ratio are favorable for carrying drugs either in the interior or on their surface, with high drug-loading capacity. In addition, abundant surface chemistry of silicon nanomaterials enables facile surface modification, facilitating the improvement of the bioavailability and targeting efficacy in the biological systems in vitro and in vivo. Moreover, silicon nanomaterials are non or lowly toxic owing to excellent biocompatibility of silicon [5–23]. Those attractive merits triggered extensive exploration of silicon nanomaterials as high-performance nanoagents for in vitro and tailorability cancer treatment, which is illustrated in this chapter.

In Sect. 5.1, we describe the development of silicon-based nanoagents for chemotherapy of cancer, typically including porous silicon (pSi)-based delivery vectors of therapeutic agents (e.g., drug molecules, proteins, and genes) and silicon nanowires (SiNWs)-based drug nanocarriers. In the following (Sect. 5.1), we then summarize the significant achievements in the fabrication of pSi- and SiNWs-based nanoagents for phototherapy applications, including photodynamic therapy and photothermal therapy of cancer. On the basis of the above two sections, future prospects are discussed in Sect. 5.3.

5.1 Chemotherapy

5.1.1 Drug Delivery

Porous silicon (pSi) is a typical kind of silicon materials containing a large number of wide pores with sizes ranging from 2 to 50 nm [11, 24]. Since the first report revealing porous silicon particles-induced enhancement of paracellular delivery of insulin [25], sufficient advancement in developing pSi-based effective drug delivery systems for various kinds of therapeutic agents (e.g., pharmaceutical drugs (indomethacin (IMC) [26, 27], doxorubicin (DOX) [28–32], paclitaxel [33], mitoxantrone dihydrochloride (MTX) [34, 35], camptothecin (CPT) [36–39], and emodin [40]), therapeutic proteins (insulin [25, 41], bovine serum albumin (BSA) [41], peptide [26, 42–46]), and genes (siRNA [33, 47–50]), etc.) has been achieved in recent decades.

In 2006, Vaccari et al. successfully developed pSi-based drug carriers for delivering DOX in a controllable manner [30]. The potential of pSi carriers in cancer therapy applications was further demonstrated via in vitro experiments. The

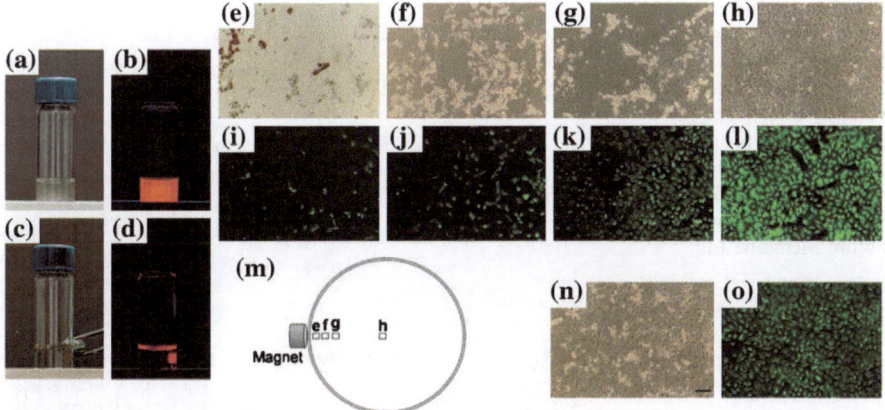

Fig. 5.1 Photos of aqueous samples of pure microparticles **a** and **b** or microparticles containing a rare earth magnet **c** and **d** irradiated by ambient light **a** and **c** or a UV light (**b** and **d**). **e–h** Microscopy pictures of DOX-loaded pSi microparticles-treated HeLa cells. **i–l** Fluorescent images of Calcein AM-treated HeLa cells. **m** The external magnet-relative position of each image. **n** and **o** Phase contrast and fluorescent pictures of the DOX-loaded pSi microparticles-treated HeLa cells imaged without magnetic guidance. Scale bars stand for 100 μm. Reproduced from Ref. [31] by permission of John Wiley and Sons Inc

results showed that this kind of Si-based carriers was in trend with the cytotoxicity of human colon adenocarcinoma cancer cells exposed to bits. In particular, the DOX-loaded pSi exhibited a continuous release behavior that reached a plateau in 5 h. Thereafter, Sailor et al. prepared nanostructured porous silicon microparticles featuring both fluorescence and magnetism via incorporating Fe_3O_4 NPs into an luminescent porous silicon substrates (Fig. 5.1) [31]. The resultant silicon-based carriers allowed high-loading capacity of DOX molecules (9.8 % by mass) and a prolonged release profile. Moreover, under the guidance of a magnetic field, drug molecules were able to be specifically delivered to HeLa cells, facilitating the decrease of DOX-induced side effect to normal tissues.

In 2012, Voelcker et al. developed a kind of biodegradable pSi-based pH responsive drug delivery system [38]. In comparison to standard CPT (the cytotoxic topoisomerase inhibitor) release behavior of uncoated pSi at different pH values, the linear release rate of CPT from the p(MAA-co-EDMA)-coated pSi matrix was largely dependent on pH values. Typically, the release rate of CPT was 13.1 nmol/(cm^2 h) at a pH value of 7.4, much faster than that (3.0 nmol/(cm^2 h)) at a pH value of 1.8. This approach is potentially suitable for delivering drug molecules that are not stable at high temperatures. The same authors also developed a kind of poly(L-lactide)-modified pSi as high-performance drug carriers, simultaneously possessing the excellent drug release properties of pSi and the favorable processability of poly(L-lactide) (Fig. 5.2) [37]. As a result, the as-prepared pSi-based drug carriers were favorable for improving drug stability and increasing the drug bioavailability at the tumor tissue. Very recently, Segal et al. employed pSi as

Fig. 5.2 **a** Schematic illustration of designing durg-loaded pSi and PLLA composite materials. **b** Cell viability of cancer cells treated by the resultant film- and monolity-based materials. Adapted from Ref. [37] with permission of Future Medicine Ltd

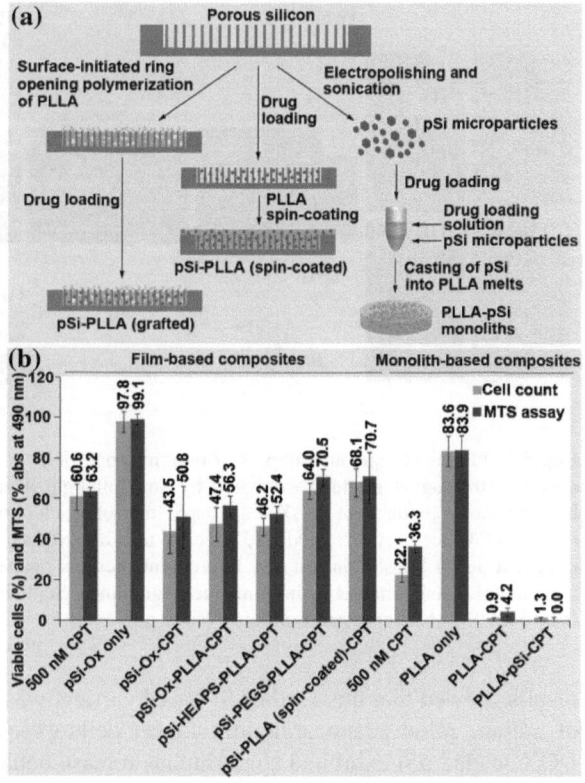

drug carriers for efficiently loading mitoxantrone dihydrochloride (MTX) [34, 35]. They showed that the modified pSi carriers exhibited persistent drug releasing behavior in a long-term manner, which was up to several weeks. Moreover, in vitro cytotoxicity experiments further showed that cancer cells (e.g., MDA-MB-231 cells) were efficiently destroyed by MTX molecules released from the prepared pSi carriers (Fig. 5.3).

Increase of circulation time in vivo and enhancement of passive targeting efficiency of drug carriers are widely regarded as two significant factors to realize high-concentration drug molecules distributed in the tumors. As a result, scientists devoted further efforts to developing pSi-based multifunctional drug delivery platform with targeting ligands and nanovalves. To date, pSi modified with specific targeting ligands (e.g., folate acid (FA) [28], peptide [43, 44], antibody [36, 51], and cyclic-RGD [52], etc.) have explored for actively targeting to cancer cells/tissues with enhanced therapeutic efficacy and minimal adverse effects. In particular, Falamaki et al. reported a smart multifunctional pSi nanocarrier modified with the FA as targeting agent, exhibiting an efficient cancer-targeted delivery and intracellular controlled release of the drug [28]. Based on systematic investigation of the drug release kinetics, they revealed that the FA molecules promoted the intracellular uptake of released DOX adjacent to the tumor cells, and further

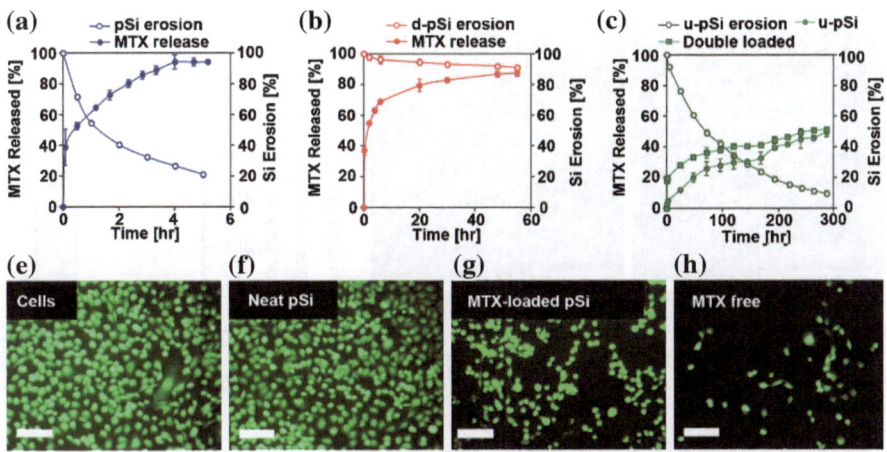

Fig. 5.3 Time-dependent released behavior of pSi erosion and MTX of the pure pSi (**a**), dodecyl-modified pSi (d-pSi, **b**) and undecanoic acid-modified pSi (u-pSi, **c**) carriers dispersed in PBS with pH value of 7.4 at 37 °C. **e–h** Fluorescent images of MDA-MB-231 cells exposed to neat pSi, MTX-loaded pSi and free MTX for 72 h. Green fluorescence is attributed to fluorescein diacetate staining (live cells). Scale bars stand for 100 μm. Reprinted from Ref. [35], Copyright 2013, with permission from Elsevier

suggested the as-prepared FA-modified pSi as a promising candidate for targeted drug delivery. Laakkonen, Santos et al. presented a kind of pSi nanovector functionalized with a tumor-homing peptide [44]. Such resultant pSi-based nanovector was highly efficacious for targeting the mammary derived growth inhibitor (MDGI) expressing cancer cells both in vitro and in vivo, thereby enhancing the accumulation of the NPs in the tumors. Specifically, after intravenous injections into nude mice bearing MDGI-expressing tumors, effective targeting was detected and the peptide-tagged pSi NPs showed ∼9-fold higher accumulation in the tumor site compared to the pure pSi NPs (Fig. 5.4). Cunin et al. presented a class of anticancer drug (camptothecin, CPT)-loaded pSiNPs, which were functionalized with cancer cell targeting antibodies [36]. In this case, semicarbazide-based bioconjugate chemistry was used to modify the pSiNP with the antibody. Moreover, the authors demonstrated the prepared pSiNPs-based carriers were capable of efficiently targeting various cancer cell lines (e.g., neuroblastoma, glioblastoma, and B lymphoma cells) via conjugation of three kinds of antibodies (e.g., MLR2, mAb528, and Rituximab), demonstrating this targeting strategy was universal to different malignant tumor models (Fig. 5.5).

On the other hand, silicon nanowires (SiNWs) have received intensive investigation since the pioneering report on synthesizing SiNWs with controllable diameters in 1998 [16, 53–56]. Of particular note, the large surface-to-volume ratio and huge-area porous structures offer exciting possibility for the design of novel SiNWs-based drug carriers with high-loading capacity. Very recently, Peng et al. for the first time employed SiNWs as high-performance drug nanovectors for delivering DOX

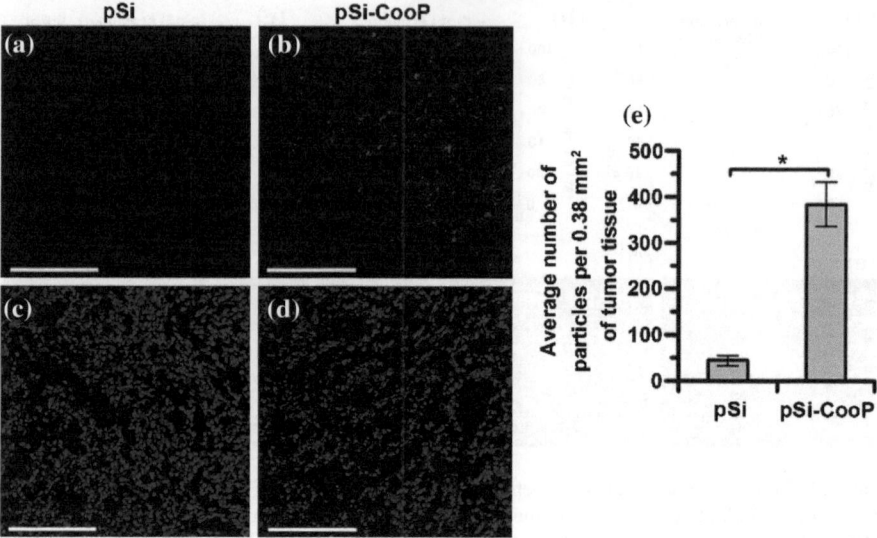

Fig. 5.4 Fluorescent photos of tumor tissue intravenously injected pSi **a** and **c** and pSi-CooP NPs (**b** and **d**). Scale bars stand for 200 μm. **e** Quantitative measurement of the average number of NPs accumulated in tumor tissue section with an area of 0.38 mm² (n ≥ 4). Reprinted from Ref. [44], Copyright 2013, with permission from Elsevier

molecules (Fig. 5.6) [21]. Of particular significance, the prepared SiNWs-based nanocarriers possess drug-loading capacity as high as ~20800 mg/g, much higher than those (1200–4000 mg/g) reported for other nanomaterial-based carriers. They further demonstrated that such SiNWs-based drug nanocarriers were highly efficacious for in vivo cancer treatment. In their experiment, tumor was efficiently inhibited, leading to long-term (e.g., 30 days) survival of the tumor-bearing mice. In sharp contrast, the tumor volume persistently increased in the control groups (e.g., tumor-bearing mice treated by pure SiNWs, free DOX, or physiological saline). This kind of SiNWs-based nanoagents is promising for widespread applications due to the facile, reproducible, and low-cost preparation of SiNWs.

5.1.2 Protein and Gene Delivery

Therapeutic proteins or genes are regarded as other kinds of therapeutic agents for cancer treatment. However, protein or gene delivery is rather difficult due to their large molecular weight and fragile structure [9]. Attributed to the porous structure, pSi-based carriers are favorable for efficient protection of proteins and genes from premature degradation. For example, Sailor et al. developed a kind of fluorescent pSi nanoparticles-based protein carriers containing multiple copies of an agonistic antibody (FGK45), which was suitable for monitoring degradation and tracking the

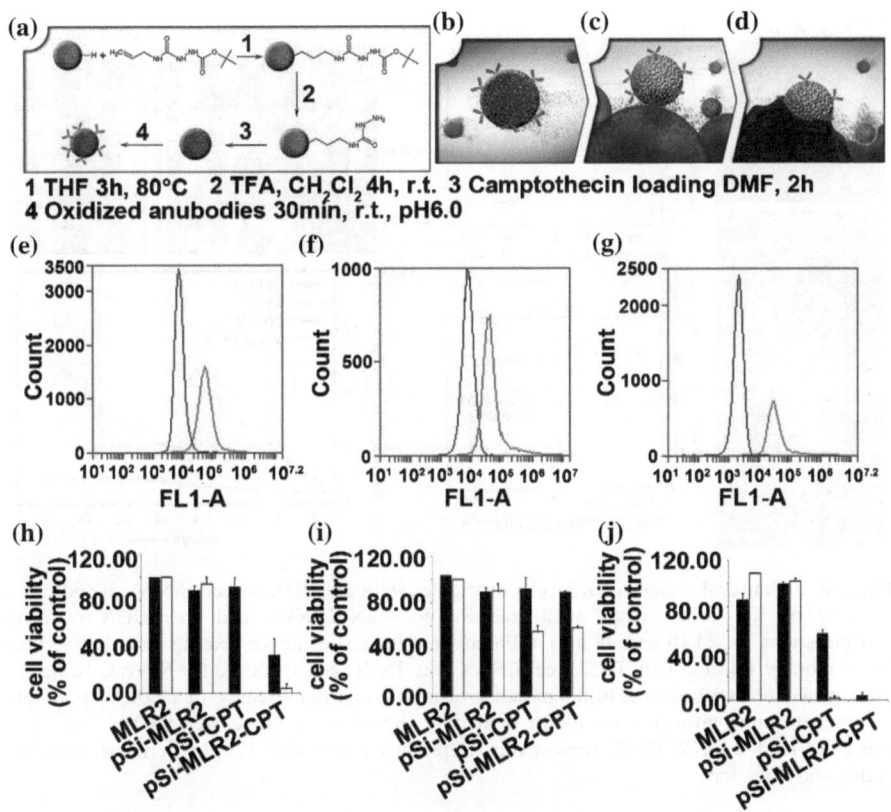

Fig. 5.5 a Schematic diagram of fabricating pSiNPs simultaneously loaded with drug molecules and modified with antibody. **b–d** Schematic illustration of pSiNPs-based vectorization of the CPT. **e–g** Flow cytometric analysis of different classes of cells (e.g., **e** SH-SY5Y/pSi-MLR2, **f** U87MG/pSi-mAb528, and **g** HRIK/pSi-Rituximab.) treated by PBS (*black*) and antibody-grafeted pSiNPs (*red*) for 30 min. **h–j** Cell viability of various cell lines (e.g., **h** SH-SY5Y cells, **i** BSR cells and **j** NSC-34 cells) treated with the pure MLR2 antibody (MLR2), MLR2 antibody-grafted pSiNPs (pSi-MLR2), CPT-loaded pSiNPs (pSi-CPT), and MLR2 antibody-grafted, CPT-loaded pSiNPs (pSi-MLR2-CPT). Reproduced from Ref. [36] by permission of John Wiley and Sons Inc

NPs inside antigen presenting cells (APCs) [57]. In 2012, Kaasalainen et al. studied the influence of medium composition (gastrointestinal peptides GLP-1 (7-37) and PYY (3-36)) on the zeta potential of pSi nanoparticles [43]. After that, Liu et al. for the first time introduced another kind of pSi-based carriers capable of co-loading a hydrophobic drug (IMC) and a hydrophilic peptide (PTT 3-36) [26]. In the same year, Kovalainen et al. prepared pSi nanoparticles loaded with a hydrophilic peptide (PYY3-36), which were workable for achieving a high peptide loading degree and sustained in vivo PYY-3-36 delivery over several days [42, 46].

Short-interfering RNA (siRNA), plasmid DNA, and antisense oligonucleotides (ASOs), serving as high-efficacy therapeutic genes, are highly attractive for cancer

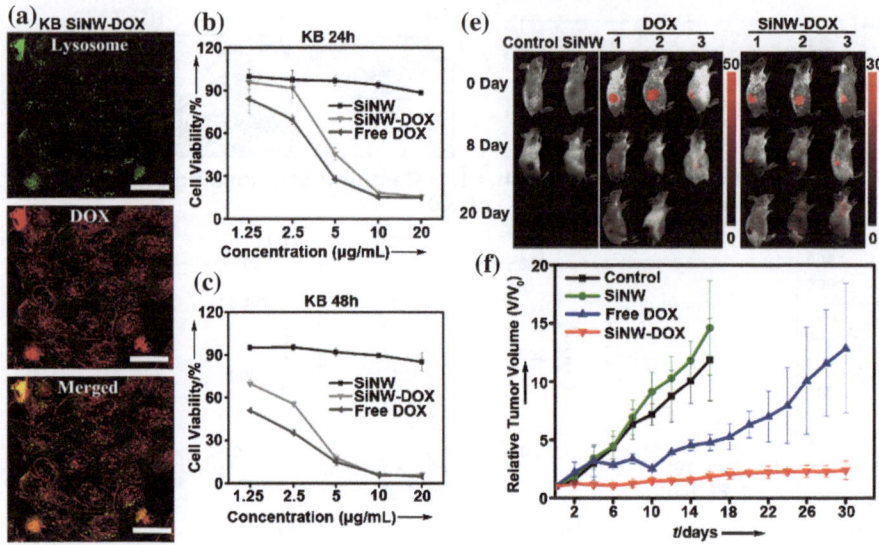

Fig. 5.6 **a** Confocal images of KB cells distributed with the DOX-loaded SiNWs. **b** and **c** Cell viability of KB cells treated with free SiNWs, SiNW-DOX, and free DOX of serial concentrations for 24 (**b** and 48 h **c**). **e** Fluorescent images of tumor-bearing nude Balb/c mice intratumorally injected with PBS, free SiNWs, free DOX, and DOX-loaded SiNWs. *White* and *red* signals are ascribed to anto-fluorescence of mouse and fluorescence of DOX, respectively. **f** Curves of tumor growth of the mouse treated by physiological saline, free SiNWs, free DOX, and DOX-loaded SiNW-DOX, respectively. Reproduced from Ref. [21] by permission of John Wiley and Sons Inc

therapy since they are much safer and low-immunogenic compared to the viral carriers [9]. One of the major challenges is lack of safe, efficient, and sustained carriers for gene loading and delivery. By utilizing porous structure and tailorable pores, pSi has been explored as high-quality carriers for gene delivery. Ferrari et al. reported a multistage vector composed of neutral nanoliposomes (dioleoyl phosphatidylcholine, DOPC)-loaded mesoporous silicon particles, which contained siRNA capable of targeting against the EphA2 oncoprotein overexpressed in a number of cancers [47]. The prepared Si-based genes carriers led to long-term (e.g., 3 weeks) sustained EphA2 gene silencing in orthotopic mouse models, suggesting the pSi-based siRNA delivery system was ready for gene silencing with wide-ranging applicability. Thereafter, Ferrari, Shen et al. employed pSi-based multistage vector (MSV) for delivery of gene-specific ataxia-telangiectasia mutated (ATM) siRNA [48]. They have demonstrated the gene-specific siRNA could efficiently knock down ATM expression, leading to highly efficient tumor growth inhibition in vivo (Fig. 5.7).

Combination therapy has been developed as a promising means for cancer therapy, in which multiple therapeutic agents featuring complementary or synergistic effect are simultaneously employed for treatment of cancer. For efficient

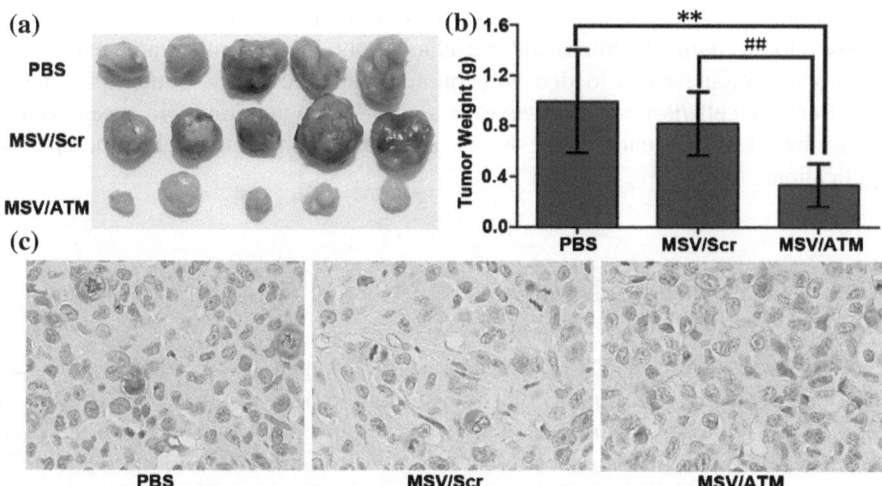

Fig. 5.7 ATM expression was knocked down through MSV/ATM treatment, resulting in efficient inhibition of tumor growth in MDA-MB-231 xenograft model. **a** Photographs of MDA-MB-231 tumor tissues treated by different reagents. **b** Quantitative measurement of tumor weight. **c** Immunohistochemical analysis of phospho-ATM expression in MDA-MB-231 tumor samples. Reproduced from Ref. [48] by permission of John Wiley and Sons Inc

combination therapy, it is of essential importance to fabricate high-performance carriers capable of concurrent delivery of various therapeutic agents. Recent investigations demonstrated pSi as promising drug co-delivery systems for cancer therapy. Typically, Shen et al. utilized pSi as a carrier to simultaneously deliver paclitaxel (or docetaxel) and EphA2 siRNA into metastatic SKOV3ip2 tumors (or chemotherapy-resistant HeyA8 ovarian tumors) [33]. Significantly, MSV/EphA2 in combination with paclitaxel (or docetaxel) led to great inhibition of tumor in the model mice. In particular, treatment of SKOV3ip2 tumor mice with MSV/EphA2 biweekly for 6 weeks resulted in dose dependent (5, 10, and 15 μg/mice) reduction of tumor weight (36 %, 64 %, and 83 %). Furthermore, combination treatment with MSV/EphA2 and docetaxel inhibited growth of HeyA8-MDR tumors, which were otherwise resistant to docetaxel treatment.

5.2 Silicon-Based Nanoagents for Phototherapy

Phototherapy (e.g., photodynamic therapy (PDT) and photothermal therapy (PTT)) is considered as another alternative manner for cancer therapy, which is capable of destroying tumor cells assisted by specific light irradiation. More importantly, in comparison to chemotherapy with limited specificity to cancer cells and undesired side effects to normal tissues/organs, nanotechnology-assisted phototherapy is more favorable for selectively destroying cancer cells and reducing side effects via controlling exposure regions of light irradiation (e.g., only the tumor lesion exposed to the light are killed) and introducing phototherapeutic nanoagents (e.g., functional nanomaterials loaded with phototherapeutic agents could specifically target cancer cells/issues via passive or active tumor targeting). In recent years, silicon-based nanoagents have shown potential promise for phototherapeutic applications.

5.2.1 Photodynamic Therapy

Photodynamic therapy (PDT) relies on reactive oxygen species (ROS) produced from photosensitizer (PS) molecules under suitable light irradiation to kill cancer cells [58]. Photosensitizers are generally organic molecules exhibiting poor selectivity for malignant tissues as well as low solubility due to aggregation in aqueous media, which is adverse to photodynamic efficiency. The incorporation of photosensitizers into nanoparticle carriers provides a means to improve bioavailability while at the same time retain the desirable photoactivity of the molecular photosensitizer. In 2009, Kuznetsov et al. investigated in vitro efficiency of the fullerene C_{60}-impregnated pSi-based photosensitizer for PDT applications [59]. Recently, Cunin et al. presented the chemical functionalization of pSiNP with a photosensitizer, by the covalent anchoring of a porphyrin into the pSi matrix using allylisocyanate-based conjugation chemistry [60]. In their experiment, successful internalization of the porphyrin-grafted pSiNP in breast cancer cells was observed, and efficient PDT from the porphyrin-pSiNP formulation was demonstrated in vitro. In 2012, Kovalev et al. further employed pSi as an efficient photosensitizer [61], serving as a promising candidate for the PDT of cancer. Sailor et al. also showed that malignant tumor cells were distinctly killed by $^1O_2$ released from the pSiNPs under irradiation of a specific light (e.g., halogen light or a light-emitting diode) (Fig. 5.8) [62]. Typically, 100 µg/mL pSiNPs resulted in $\sim$45 % cell deaths of HeLa or NIH-3T3 cells exposed to 60 J/cm^2 white light for 10 min. In contrary, HeLa or NIH-3T3 cells without pSiNPs treatment preserved $\sim$90 % or 75 % cell viability, respectively. This study provides the first example of pSiNPs-assisted photodynamic destruction of malignant tumor cells, suggesting pSi as a potentially nontoxic, biodegradable alternative to molecular PDT agents widely used in clinic applications up to present. Since then, lots of works concerning pSiNPs-based PDT of cancer have been carried out. For example, Osminkina et al. demonstrated that aqueous suspensions of silicon nanocrystals prepared from pSi could act as photosensitizers and sono-sensitizers for cancer therapy [63]. They showed that combined action of silicon nanocrystals and ultrasound irradiation led to the destruction of NIH-3T3 and Hep-2 cells, suggesting possibilities of the theranostic applications of silicon nanocrystals for cancer treatment. Gonzalez et al. demonstrated that small-sized (<5 nm) SiNPs facilitated enhancement of yielding $O_2^{\bullet-}/HO_2^{\bullet}$ and $HO^{\bullet}$ in aqueous phase, which may serve as novel radio-sensitizes for cancer radiotherapy with improved therapeutic outcomes [64].

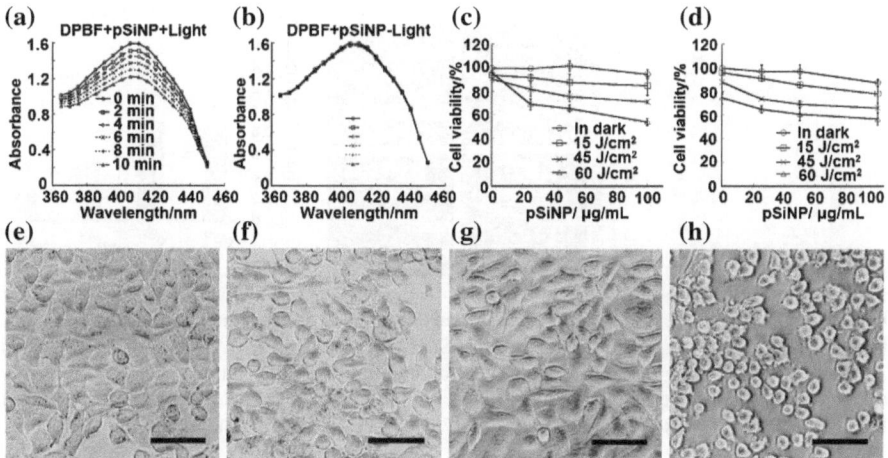

Fig. 5.8 Absorption spectra of the singlet oxygen indicator DPBF with pSiNP irradiated by light **a** or without light (**b**). Cell viability of pSiNPs-treated HeLa **c** and NIH-3T3 **d** cells. Phase contrast microscope pictures of PBS (**e**, in the *dark*)-, pSiNP (**f**, in the *dark* for 10 min)-, PBS (**g**, irradiated by 60 J/cm² light for 10 min)-, and pSiNP (**h**, irradiated by 60 J/cm² light for 10 min)-treated HeLa cells. Scale bars stand for 50 μm. Reprinted with the permission from Ref. [62]. Copyright 2011 American Chemical Society

5.2.2 Photothermal Therapy

For photothermal therapy (PTT), an optical-absorbing agent is required to generate heat under light irradiation, producing a local high temperature to kill cancer cells. In the past decade, numerous classes of nanomaterials with strong near-infrared (NIR) absorbance have been explored as a photothermal agent for cancer PTT treatment, including various gold nanostructures, carbon nanomaterials, Pd nanosheets, CuS nanoparticles, and organic nanoparticles (e.g., poly-(3,4-ethylenedioxythiophene):poly(4-styrenesulfonate) (PEDOT:PSS) nanoparticles) [65–74]. In 2007, Lee et al. utilized pSi as a therapeutic agent, which generated sufficient heat to kill cancer cells [75]. It is worthwhile to point out that, while the surface temperature of pSi increased as high as that of carbon nanotubes (CNT), pSi produced a smaller amount of ROS than CNT during NIR light irradiation. In vitro results suggested that cancer cells could be selectively destroyed by the pSi-based nanoagents assisted by NIR light irradiation, with undetectable side effects on the surrounding healthy cells [76]. In the following study, based on comparison of photothermal properties and the cytotoxic effect, they further revealed that pSiO exhibited lower photothermal properties and cell death rate compared to bare pSi [77]. Meanwhile, Shen et al. explored pSi nanoassembly with hollow gold nanoshells for photothermal therapeutics (Fig. 5.9) [78]. Compared to free AuNPs, the resultant pSi/HAuNSs nanoassembly showed higher efficiency of heat generation, which was thus highly efficacious for treatment of human and mouse breast cancer

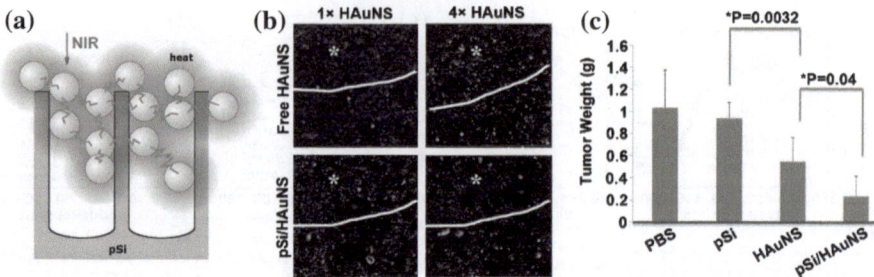

Fig. 5.9 **a** Schematic representation of pSi/HAuNSs-induced enhancement of thermal efficiency.
b Cell viability staining of free HAuNSs- or pSi/HAuNSs-treated MDA-MB-231 cells. Calcein
AM (*green*) and EthD-1 (*red*) were employed for staining live cells and dead cells, respectively.
c Tumor weight measurement of murine 4T1 tumor treated by PBS, pSi, HAuNS, and Psi/
HAuNS, respectively. Reproduced from Ref. [78] by permission of John Wiley and Sons Inc

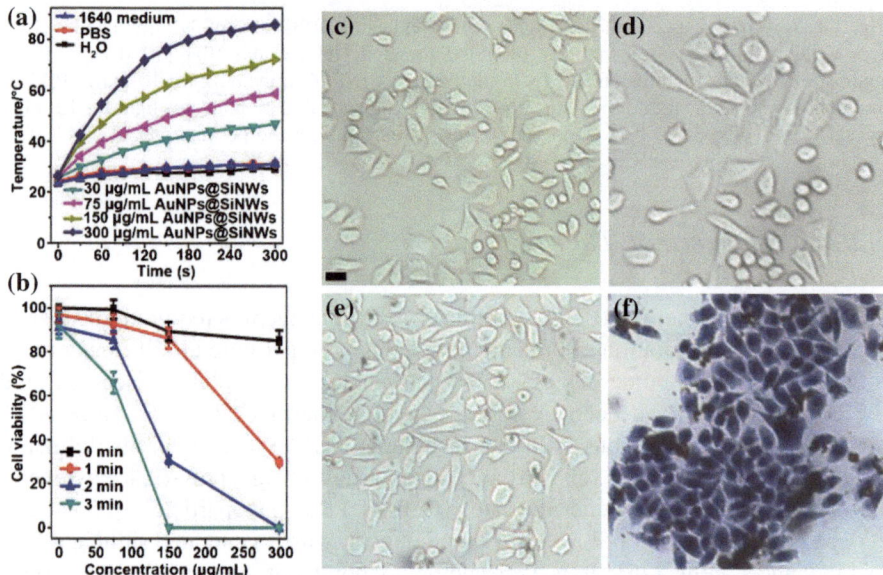

Fig. 5.10 **a** Temperature increase of AuNPs@SiNWs aqueous solutions with serial concentrations under NIR irradiation (2 W/cm^2) for different time. **b** Cell viability of KB cells treated by
the AuNPs@SiNWs irradiated by NIR light for various irradiation time ranging from 0 to 3 min.
c–f microscopy pictures of KB cells without any treatment (**c**), under NIR irradiation (**d**), treated
with the PEG-AuNPs@SiNWs (**e**), and treated with the PEG-AuNPs@SiNWs under 3-min NIR
irradiation (**f**). Dead cells were stained with Trypan *blue*. Scale bars stand for 20 µm. Reprinted
with the permission from Ref. [20]. Copyright 2012 American Chemical Society

lines in vitro and the mouse model of 4T1 mammary tumor in vivo. By utilizing
strong NIR absorbance of SiNWs, Su et al. recently developed a kind of SiNWs-
based hyperthermia agents made of AuNPs-decorated SiNWs for photothermal
cancer therapy [20]. In this case, AuNPs and SiNWs acted as heat producer and

NIR trapper, respectively. As a result, the AuNPs@SiNWs, serving as high-quality NIR hyperthermia agent, could rapidly generate high heat under NIR irradiation for destruction of cancer cells. They further demonstrated that the SiNWs-based nanoagents were universal for treatment of other cancer cells lines (e.g., A549 and HeLa cells were completely destructed by the AuNPs@SiNWs, as shown in Fig. 5.10). Meanwhile, Park et al. reported gold nanoclusters (AuNCs)-coated SiNWs as a substrate for concurrent capture and phototherapy of tumor cells [68]. In their work, to capture tumor cells with high specificity, specific antibody was first conjugated with the AuNC-coated SiNWs. Afterward, the captured cancer cells were efficiently killed by NIR irradiation due to sufficient heat produced by the AuNPs@SiNWs. Consequently, such SiNWs-based platform may provide novel strategies for simultaneously capturing and destroying tumor cells.

5.3 Conclusions and Perspectives

We have reviewed representative advancement of silicon-based nanoagents for cancer therapy in this chapter. Typically, taking advantage of large surface-to-volume ratio, porous structures and surface tailorability, silicon materials (e.g., pSi and SiNWs) have been successfully explored as high-performance drug molecules, therapeutic proteins, and genes carriers, which are highly efficacious for cancer chemotherapy. Moreover, silicon-based nanoagents hold great promise for photodynamic and photothermal treatment of cancer, completely destroying tumor cells in vitro and greatly inhibiting tumor growth in vivo assisted by specific light irradiation. Despite such exciting progresses, major challenges are still remaining including development of multifunctional silicon-based nanogents featuring unique optical/magnetic/thermal properties, to meet increasing demand of multi-model cancer treatments. Besides, there currently exists scant information on in vitro and in vivo behaviors of the silicon-based nanoagents, which are the preconditions for their wide-ranging applications and thus needed to be systematically investigated in the future.

References

1. Allen TM, Cullis PR (2004) Drug delivery systems: entering the mainstream. Science 303(5665):1818–1822
2. Iyer AK, Singh A, Ganta S, Amiji MM (2013) Role of integrated cancer nanomedicine in overcoming drug resistance. Adv Drug Deliver Rev 65(13–14):1784–1802
3. Yang K, Feng L, Shi X, Liu Z (2013) Nano-graphene in biomedicine: theranostic applications. Chem Soc Rev 42(2):530–547
4. Hubbell JA, Chilkoti A (2012) Nanomaterials for drug delivery. Science 337(6092):303–305

5. Jarvis KL, Barnes TJ, Prestidge CA (2012) Surface chemistry of porous silicon and implications for drug encapsulation and delivery applications. Adv Colloid Interface Sci 175:25–38
6. Jaganathan H, Godin B (2012) Biocompatibility assessment of Si-based nano- and microparticles. Adv Drug Deliver Rev 64(15):1800–1819
7. Fine D, Grattoni A, Goodall R, Bansal SS, Chiappini C, Hosali S, van de Ven AL, Srinivasan S, Liu X, Godin B, Brousseau L, Yazdi IK, Fernandez-Moure J, Tasciotti E, Wu H-J, Hu Y, Klemm S, Ferrari M (2013) Silicon micro- and nanofabrication for medicine. Adv Health Mater 2(5):632–666
8. Li B-R, Chen C-W, Yang W-L, Lin T-Y, Pan C-Y, Chen Y-T (2013) Biomolecular recognition with a sensitivity-enhanced nanowire transistor biosensor. Biosens Bioelectron 45:252–259
9. Tang F, Li L, Chen D (2012) Mesoporous silica nanoparticles: synthesis, biocompatibility and drug delivery. Adv Mater 24(12):1504–1534
10. Shahbazi M-A, Herranz B, Santos HA (2012) Nanostructured porous Si-based nanoparticles for targeted drug delivery. Biomatter 2(4):296–312
11. Barnes TJ, Jarvis KL, Prestidge CA (2013) Recent advances in porous silicon technology for drug delivery. Ther Deliv 4(7):811–823
12. Haidary SM, Córcoles EP, Ali NK (2012) Nanoporous silicon as drug delivery systems for cancer therapies. J Nanomater (18), 830503
13. Salonen J, Kaukonen AM, Hirvonen J, Lehto V-P (2008) Mesoporous silicon in drug delivery applications. J Pharm Sci 97(2):632–653
14. Bonanno LM, Segal E (2011) Nanostructured porous silicon-polymer-based hybrids: from biosensing to drug delivery. Nanomedicine 6(10):1755–1770
15. He Y, Su Y, Yang X, Kang Z, Xu T, Zhang R, Fan CH, Lee S-T (2009) Photo and pH stable, highly-luminescent silicon nanospheres and their bioconjugates for immunofluorescent cell imaging. J Am Chem Soc 131(12):4434–4438
16. He Y, Fan CH, Lee S-T (2010) Silicon nanostructures for bioapplications. Nano Today 5(4):282–295
17. He Y, Su S, Xu T, Zhong Y, Zapien JA, Li J, Fan CH, Lee S-T (2011) Silicon nanowires-based highly-efficient SERS-active platform for ultrasensitive DNA detection. Nano Today 6(2):122–130
18. Jiang Z, Jiang X, Su S, Wei X, Lee S-T, He Y (2012) Silicon-based reproducible and active surface-enhanced Raman scattering substrates for sensitive, specific, and multiplex DNA detection. Appl Phys Lett 100(20):203104
19. Su S, Wei X, Zhong Y, Guo Y, Su Y, Huang Q, Lee S-T, Fan C, He Y (2012) Silicon nanowire-based molecular beacons for high-sensitivity and sequence-specific DNA multiplexed analysis. ACS Nano 6(3):2582–2590
20. Su Y, Wei X, Peng F, Zhong Y, Lu Y, Su S, Xu T, Lee S-T, He Y (2012) Gold nanoparticles-decorated silicon nanowires as highly efficient near-infrared hyperthermia agents for cancer cells destruction. Nano Lett 12(4):1845–1850
21. Peng F, Su Y, Wei X, Lu Y, Zhou Y, Zhong Y, Lee S-T, He Y (2013) Silicon-nanowire-based nanocarriers with ultrahigh drug-loading capacity for in vitro and in vivo cancer therapy. Angew Chem Int Ed 52(5):1457–1461
22. Wei X, Su S, Guo Y, Jiang X, Zhong Y, Su Y, Fan C, Lee S-T, He Y (2013) A molecular beacon-based signal-off surface-enhanced Raman scattering strategy for highly sensitive, reproducible, and multiplexed DNA detection. Small 9(15):2493–2499
23. Park J-H, Gu L, von Maltzahn G, Ruoslahti E, Bhatia SN, Sailor MJ (2009) Biodegradable luminescent porous silicon nanoparticles for in vivo applications. Nat Mater 8(4):331–336
24. Prestidge CA, Barnes TJ, Lau C-H, Barnett C, Loni A, Canham L (2007) Mesoporous silicon: a platform for the delivery of therapeutics. Exp Opin Drug Deliv 4(2):101–110
25. Foraker A, Walczak R, Cohen M, Boiarski T, Grove C, Swaan P (2003) Microfabricated porous silicon particles enhance paracellular delivery of insulin across intestinal Caco-2 cell monolayers. Pharm Res 20(1):110–116

26. Liu D, Bimbo LM, Mäkilä E, Villanova F, Kaasalainen M, Herranz-Blanco B, Caramella CM, Lehto V-P, Salonen J, Herzig K-H, Hirvonen J, Santos HA (2013) Co-delivery of a hydrophobic small molecule and a hydrophilic peptide by porous silicon nanoparticles. J Control Release 170(2):268–278

27. Bimbo LM, Sarparanta M, Makila E, Laaksonen T, Laaksonen P, Salonen J, Linder MB, Hirvonen J, Airaksinen AJ, Santos HA (2012) Cellular interactions of surface modified nanoporous silicon particles. Nanoscale 4(10):3184–3192

28. Tabasi O, Falamaki C, Khalaj Z (2012) Functionalized mesoporous silicon for targeted-drug-delivery. Colloid Surface B 98:18–25

29. Xu Z, Wang D, Guan M, Liu X, Yang Y, Wei D, Zhao C, Zhang H (2012) Photoluminescent silicon nanocrystal-based multifunctional carrier for pH-regulated drug delivery. ACS Appl Mater Interfaces 4(7):3424–3431

30. Vaccari L, Canton D, Zaffaroni N, Villa R, Tormen M, di Fabrizio E (2006) Porous silicon as drug carrier for controlled delivery of doxorubicin anticancer agent. Microelectron Enginee 83(4–9):1598–1601

31. Gu L, Park J-H, Duong KH, Ruoslahti E, Sailor MJ (2010) Magnetic luminescent porous silicon microparticles for localized delivery of molecular drug payloads. Small 6(22):2546–2552

32. Wu EC, Park J-H, Park J, Segal E, Cunin F, Sailor MJ (2008) Oxidation-triggered release of fluorescent molecules or drugs from mesoporous Si microparticles. ACS Nano 2(11):2401–2409

33. Shen H, Rodriguez-Aguayo C, Xu R, Gonzalez-Villasana V, Mai J, Huang Y, Zhang G, Guo X, Bai L, Qin G, Deng X, Li Q, Erm DR, Aslan B, Liu X, Sakamoto J, Chavez-Reyes A, Han H-D, Sood AK, Ferrari M, Lopez-Berestein G (2013) Enhancing chemotherapy response with sustained EphA2 silencing using multistage vector delivery. Clin Cancer Res 19(7):1806–1815

34. Zilony N, Tzur-Balter A, Segal E, Shefi O (2013) Bombarding cancer: biolistic delivery of therapeutics using porous Si carriers. Sci Rep 3. doi:10.1038/srep02499

35. Tzur-Balter A, Gilert A, Massad-Ivanir N, Segal E (2013) Engineering porous silicon nanostructures as tunable carriers for mitoxantrone dihydrochloride. Acta Biomater 9(4):6208–6217

36. Secret E, Smith K, Dubljevic V, Moore E, Macardle P, Delalat B, Rogers M-L, Johns TG, Durand J-O, Cunin F, Voelcker NH (2013) Antibody-functionalized porous silicon nanoparticles for vectorization of hydrophobic drugs. Adv Health Mater 2(5):718–727

37. McInnes SJ, Irani Y, Williams KA, Voelcker NH (2012) Controlled drug delivery from composites of nanostructured porous silicon and poly (L-lactide). Nanomedicine 7(7):995–1016

38. McInnes SJP, Szili EJ, Al-Bataineh SA, Xu J, Alf ME, Gleason KK, Short RD, Voelcker NH (2012) Combination of iCVD and porous silicon for the development of a controlled drug delivery system. ACS Appl Mater Interfaces 4(7):3566–3574

39. Botella P, Ortega I, Quesada M, Madrigal RF, Muniesa C, Fimia A, Fernandez E, Corma A (2012) Multifunctional hybrid materials for combined photo and chemotherapy of cancer. Dalton Trans 41(31):9286–9296

40. Hernandez M, Recio G, Martin-Palma RJ, Garcia-Ramos JV, Domingo C, Sevilla P (2012) Surface enhanced fluorescence of anti-tumoral drug emodin adsorbed on silver nanoparticles and loaded on porous silicon. Nanoscale Res Lett 7, 364(1):1–7

41. Pastor E, Matveeva E, Valle-Gallego A, Goycoolea FM, Garcia-Fuentes M (2011) Protein delivery based on uncoated and chitosan-coated mesoporous silicon microparticles. Colloid Surface B 88(2):601–609

42. Kovalainen M, Mönkäre J, Mäkilä E, Salonen J, Lehto V-P, Herzig K-H, Järvinen K (2012) Mesoporous silicon (PSi) for sustained peptide delivery: effect of pSi microparticle surface chemistry on peptide YY3-36 release. Pharm Res 29(3):837–846

43. Kaasalainen M, Mäkilä E, Riikonen J, Kovalainen M, Järvinen K, Herzig K-H, Lehto V-P, Salonen J (2012) Effect of isotonic solutions and peptide adsorption on zeta potential of porous silicon nanoparticle drug delivery formulations. Int J Pharmaceut 431(1–2):230–236
44. Kinnari PJ, Hyvönen MLK, Mäkilä EM, Kaasalainen MH, Rivinoja A, Salonen JJ, Hirvonen JT, Laakkonen PM, Santos HA (2013) Tumour homing peptide-functionalized porous silicon nanovectors for cancer therapy. Biomaterials 34(36):9134–9141
45. Huotari A, Xu W, Mönkäre J, Kovalainen M, Herzig K-H, Lehto V-P, Järvinen K (2013) Effect of surface chemistry of porous silicon microparticles on glucagon-like peptide-1 (GLP-1) loading, release and biological activity. Int J Pharmaceut 454(1):67–73
46. Kovalainen M, Mönkäre J, Kaasalainen M, Riikonen J, Lehto V-P, Salonen J, Herzig K-H, Järvinen K (2013) Development of porous silicon nanocarriers for parenteral peptide delivery. Mol Pharmaceutics 10(1):353–359
47. Tanaka T, Mangala LS, Vivas-Mejia PE, Nieves-Alicea R, Mann AP, Mora E, Han H-D, Shahzad MM, Liu X, Bhavane R, Gu J, Fakhoury JR, Chiappini C, Lu C, Matsuo K, Godin B, Stone RL, Nick AM, Lopez-Berestein G, Sood AK, Ferrari M (2010) Sustained small interfering RNA delivery by mesoporous silicon particles. Cancer Res 70(9):3687–3696
48. Xu R, Huang Y, Mai J, Zhang G, Guo X, Xia X, Koay EJ, Qin G, Erm DR, Li Q, Liu X, Ferrari M, Shen H (2013) Multistage vectored siRNA targeting ataxia-telangiectasia mutated for breast cancer therapy. Small 9(9–10):1799–1808
49. Shen J, Xu R, Mai J, Kim H-C, Guo X, Qin G, Yang Y, Wolfram J, Mu C, Xia X, Gu J, Liu X, Mao Z-W, Ferrari M, Shen H (2013) High capacity nanoporous silicon carrier for systemic delivery of gene silencing therapeutics. ACS Nano 7(11):9867–9880
50. Ferrari M (2010) Experimental therapies: Vectoring siRNA therapeutics into the clinic. Nat Rev Clin Oncol 7(9):485–486
51. Yokoi K, Godin B, Oborn CJ, Alexander JF, Liu X, Fidler IJ, Ferrari M (2013) Porous silicon nanocarriers for dual targeting tumor associated endothelial cells and macrophages in stroma of orthotopic human pancreatic cancers. Cancer Lett 334(2):319–327
52. Van De Ven AL, Kim P, Haley OH, Fakhoury JR, Adriani G, Schmulen J, Moloney P, Hussain F, Ferrari M, Liu X, Yun S-H, Decuzzi P (2012) Rapid tumoritropic accumulation of systemically injected plateloid particles and their biodistribution. J Control Release 158(1):148–155
53. Zhang YF, Tang YH, Wang N, Yu DP, Lee CS, Bello I, Lee S-T (1998) Silicon nanowires prepared by laser ablation at high temperature. Appl Phys Lett 72(15):1835–1837
54. Morales AM, Lieber CM (1998) A laser ablation method for the synthesis of crystalline semiconductor nanowires. Science 279(5348):208–211. doi:10.1126/science.279.5348.208
55. Wang Y, Wang T, Da P, Xu M, Wu H, Zheng G (2013) Silicon nanowires for biosensing, energy storage, and conversion. Adv Mater 25(37):5177–5195
56. Liu Z, Sun X, Nakayama-Ratchford N, Dai H (2007) Supramolecular chemistry on water-soluble carbon nanotubes for drug loading and delivery. ACS Nano 1(1):50–56
57. Gu L, Ruff LE, Qin Z, Corr M, Hedrick SM, Sailor MJ (2012) Multivalent porous silicon nanoparticles enhance the immune activation potency of agonistic CD40 antibody. Adv Mater 24(29):3981–3987
58. Dolmans DEJGJ, Fukumura D, Jain RK (2003) Photodynamic therapy for cancer. Nat Rev Cancer 3(5):380–387
59. Kuznetsov SN, Volkova TO, Pikulev VB, Saren AA, Gardin YE, Gurtov VA (2009) Porous silicon versus its composite with C60: testing in vitro. Phys Status Solidi A 206(6):1352–1355
60. Secret E, Maynadier M, Gallud A, Gary-Bobo M, Chaix A, Belamie E, Maillard P, Sailor MJ, Garcia M, Durand J-O, Cunin F (2013) Anionic porphyrin-grafted porous silicon nanoparticles for photodynamic therapy. Chem Commun 49(39):4202–4204
61. Kovalev D, Gross E, Künzner N, Koch F, Timoshenko VY, Fujii M (2002) Resonant electronic energy transfer from excitons confined in silicon nanocrystals to oxygen molecules. Phys Rev Lett 89(13):137401

62. Xiao L, Gu L, Howell SB, Sailor MJ (2011) Porous silicon nanoparticle photosensitizers for singlet oxygen and their phototoxicity against cancer cells. ACS Nano 5(5):3651–3659
63. Osminkina LA, Gongalsky MB, Motuzuk AV, Timoshenko VY, Kudryavtsev AA (2011) Silicon nanocrystals as photo- and sono-sensitizers for biomedical applications. Appl Phys B 105(3):665–668
64. David Gara P, Garabano N, Llansola Portoles M, Moreno MS, Dodat D, Casas O, Gonzalez M, Kotler M (2012) ROS enhancement by silicon nanoparticles in X-ray irradiated aqueous suspensions and in glioma C6 cells. J Nanopart Res 14(3):1–13
65. Yavuz MS, Cheng Y, Chen J, Cobley CM, Zhang Q, Rycenga M, Xie J, Kim C, Song KH, Schwartz AG, Wang LV, Xia Y (2009) Gold nanocages covered by smart polymers for controlled release with near-infrared light. Nat Mater 8(12):935–939
66. Xia Y, Li W, Cobley CM, Chen J, Xia X, Zhang Q, Yang M, Cho EC, Brown PK (2011) Gold nanocages: from synthesis to theranostic applications. Acc Chem Res 44(10):914–924
67. Loo C, Lowery A, Halas N, West J, Drezek R (2005) Immunotargeted nanoshells for integrated cancer imaging and therapy. Nano Lett 5(4):709–711. doi:10.1021/nl050127s
68. Park G-S, Kwon H, Kwak DW, Park SY, Kim M, Lee J-H, Han H, Heo S, Li XS, Lee JH, Kim YH, Lee J-G, Yang W, Cho HY, Kim SK, Kim K (2012) Full surface embedding of gold clusters on silicon nanowires for efficient capture and photothermal therapy of circulating tumor cells. Nano Lett 12(3):1638–1642
69. Chakravarty P, Marches R, Zimmerman NS, Swafford AD-E, Bajaj P, Musselman IH, Pantano P, Draper RK, Vitetta ES (2008) Thermal ablation of tumor cells with antibody-functionalized single-walled carbon nanotubes. Proc Natl Acad Sci USA 105(25):8697–8702
70. Hirsch LR, Stafford RJ, Bankson JA, Sershen SR, Rivera B, Price RE, Hazle JD, Halas NJ, West JL (2003) Nanoshell-mediated near-infrared thermal therapy of tumors under magnetic resonance guidance. Proc Natl Acad Sci USA 100(23):13549–13554
71. Kam NWS, O'Connell M, Wisdom JA, Dai H (2005) Carbon nanotubes as multifunctional biological transporters and near-infrared agents for selective cancer cell destruction. Proc Natl Acad Sci USA 102(33):11600–11605
72. Huang X, Tang S, Mu X, Dai Y, Chen G, Zhou Z, Ruan F, Yang Z, Zheng N (2011) Freestanding palladium nanosheets with plasmonic and catalytic properties. Nat Nanotecnol 6(1):28–32
73. Tian Q, Jiang F, Zou R, Liu Q, Chen Z, Zhu M, Yang S, Wang J, Wang J, Hu J (2011) Hydrophilic Cu9S5 nanocrystals: A photothermal agent with a 25.7 % heat conversion efficiency for photothermal ablation of cancer cells in vivo. ACS Nano 5(12):9761–9771
74. Cheng L, Yang K, Chen Q, Liu Z (2012) Organic stealth nanoparticles for highly effective in vivo near-infrared photothermal therapy of cancer. ACS Nano 6(6):5605–5613
75. Lee C, Kim H, Cho Y, Lee WI (2007) The properties of porous silicon as a therapeutic agent via the new photodynamic therapy. J Mater Chem 17(25):2648–2653
76. Lee C, Kim H, Hong C, Kim M, Hong SS, Lee DH, Lee WI (2008) Porous silicon as an agent for cancer thermotherapy based on near-infrared light irradiation. J Mater Chem 18(40):4790–4795
77. Lee C, Hong C, Lee J, Son M, Hong S-S (2012) Comparison of oxidized porous silicon with bare porous silicon as a photothermal agent for cancer cell destruction based on in vitro cell test results. Laser Med Sci 27(5):1001–1008
78. Shen H, You J, Zhang G, Ziemys A, Li Q, Bai L, Deng X, Erm DR, Liu X, Li C, Ferrari M (2012) Cooperative, nanoparticle-enabled thermal therapy of breast cancer. Adv Health Mater 1(1):84–89

Chapter 6
Biosafety Assessment of Silicon Nanomaterials

Abstract Biosafety assessment of nanomaterials is of essential importance and regarded as a critical precondition for practical applications. Silicon nanomaterials have shown great promise for myriad biological and biomedical applications, including biosensing, bioimaging, cancer diagnosis and therapy, etc. It is worthwhile to point out that, while favorable biocompatibility is accredited to silicon, systematic and reliable biosafety assessment of silicon nanomaterials is required to be carried out for practical applications. Scientists have made pioneer work to investigate in vitro and in vivo behaviors (e.g., cellular viability, biodistribution, pharmacokinetics, etc.) of two typical kinds of silicon nanomaterials (i.e., silicon nanoparticles (SiNPs) and silicon nanowires (SiNWs)). These primary results suggest favorable biocompatibility of silicon nanomaterials in general; notwithstanding, potential safety concerns are also addressed by several studies, based on investigation of detectable in vitro and in vivo toxicity induced by silicon nanomaterials. This chapter intends to take SiNPs and SiNWs as models and discuss topics concerning silicon materials-related biosafety investigation, with the hope to outline these pioneer studies as the starting points for risk assessment of silicon nanostructures.

Keywords Biosafty assessment · Biocompatibility · Cytotoxicity · In vitro and in vivo · Silicon nanoparticles · Silicon nanowire

Nanoscience has matured significantly as it has been transferred from the bench top to the bedside. Nowadays, nanomaterials have been used in a wide variety of commercial products [1]. With the increasing environmental and human exposure to nanomaterials, the concept of nanotoxicology emerges [2]. For example, it was found that nanometer-sized titanium dioxide particles used in sunscreens could induce brain damage in mice [3]. Besides, despite favorable biocompatibility of bulk carbon, carbon nanotubes may produce asbestos-like effects on cells and soot particles are potentially adverse to human health [4].

For nanomaterials-based biomedical applications involving deliberate, direct ingestion or injection of nanomaterials into the body [5–7], it will be more critical to consider the toxicity. Till now, the toxic effects of various nanomaterials (e.g.,

Y. He and Y. Su, *Silicon Nano-biotechnology*, SpringerBriefs in Molecular Science, DOI: 10.1007/978-3-642-54668-6_6, © The Author(s) 2014

metal-and carbon-based nanomaterials) have been extensively investigated, which have been concluded in several literature reviews [1, 5, 8–10]. As described in previous chapters, while silicon-based nanomaterials have shown great promise for myriad biological and biomedical applications due to low inherent toxicity of silicon, there is still remaining a major challenge for systematic in vitro and in vivo safety assessment of silicon nanomaterials in a reliable manner. As a consequence, research on the biological effects of silicon nanomaterials has attracted intensive attentions. In this chapter, we summarize the primary studies in investigating the factors affecting biocompatibility of SiNPs and SiNWs.

In Sect. 6.1, we describe dominant achievement in biosafety assessment of SiNPs, revealing the relationship between in vitro and in vivo toxicity and chemical/physical properties of SiNPs (e.g., morphologies, surface ligands, etc.). SiNWs-related biosafety investigation is illustrated in Sect. 6.2. In the final section, i.e., Sect. 6.3, we discuss challenges and perspectives for silicon nanostructures-based biosafety assessment.

6.1 SiNPs-Related Biosafety Assessment

Fluorescent silicon nanoparticles (SiNPs) have gained great attentions due to their unique optical properties. Moreover, the potential to provide intrinsically nontoxic NPs makes SiNPs as highly promising bioprobes. However, though they have a non or lowly toxic reputation in bulk and molecular forms, the properties of silicon may probably change on the nanoscale [11–13]; as a consequence, risk assessment of nanoscale silicon is demanded for drawing firm conclusion. Up till now, knowledge of the biosafety of fluorescent SiNPs is limited, although some studies show promising toxicity profiles [14–28].

During the past several years, the cytotoxicity of SiNPs synthesized through different methods has been basically assessed via measurement of cell viability using an established MTT [3-(4,5-dimethylthiazol-2-yl)-2,5-diphenyltetrazolium bromide] assay and/or observing the change of cell morphology [16, 21, 26, 29]. In 2010, Tilley et al. demonstrated that the 50 % inhibition coefficients (IC_{50}) for WS1 cells and A549 cells were 0.068 and 0.13 mg/mL, respectively, when cells were incubated with epoxides-terminated SiNPs [16]. In contrast, diol-terminated SiNPs featured higher IC50 values (IC_{50} for WS1 cells and A549 cells were 0.097 and 0.225 mg/mL, respectively), suggesting that surface properties of SiNPs largely influenced toxicity [16]. Whereas, Chao and coworkers showed that the amine-terminated SiNPs prepared from electrochemically etched porous silicon did not remarkably affect the proliferation of HepG2 cells even when the concentration reached 0.2 mg/mL and the incubation time increased to 48 h [26]. Our group also demonstrated that the SiNPs synthesized through microwave-assisted strategies possessed favorable biocompatibility, e.g., Hela cells incubated with the prepared SiNPs preserved high cell viability (<90 %) for 48 h [29].

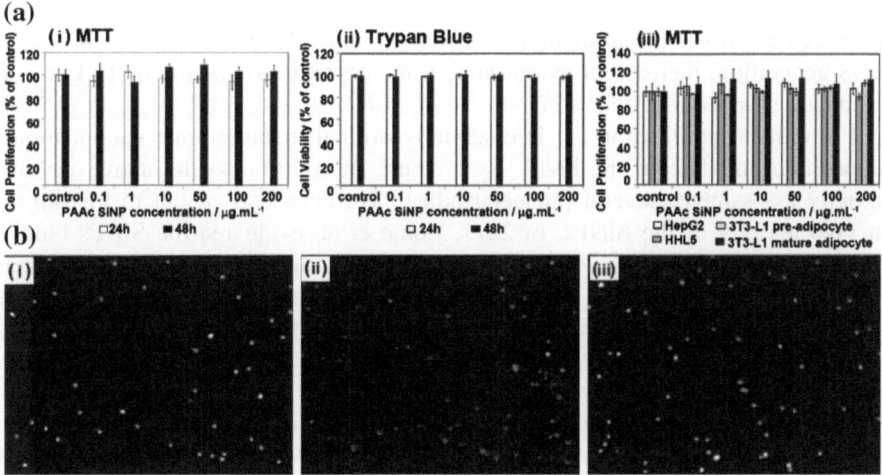

Fig. 6.1 a Cell proliferation and viability of HepG2 cells exposed to the PAA-SiNPs with different concentrations ranging from 0.1 to 200 µg/mL for 24 and 48 h, determined by well-established MTT assay (*i*) and typan blue staining (*ii*), respectively. (*iii*) cell proliferation of various cell lines (e.g., HepG2, HHL5, 3T3-L1 pre-adipocyte, and 3T3-L1 mature adipocyte) treated with the PAA-SiNPs (0.1–200 µg/mL) for 48 h, determined by MTT assay. **b** Comet IV software-assisted analysis of the SiNPs-induced DNA damage in HepG2 cells. (*i–iii*) cells were treated with medium only (*i*), 50 µM H$_2$O$_2$ for 1 h (*ii*), or 100 µg/mL PAA-SiNPs for 1 h (*iii*), respectively. Reproduced from ref. [22] by permission of John Wiley & Sons Inc

In 2009, the cytotoxicity of nanosized silicon particle (3 nm in diameter) was compared with silicon micro-sized particles ($\sim$100–3,000 nm in diameter) in RAW 264.7 macrophages through evaluating cell viability and inflammatory responses [15]. It was found that low-concentration ($\leq$20 µg/mL) SiNPs produced feeble cytotoxicity or inflammatory responses during 48-h incubation. In comparison, SiNPs whose concentration was larger than 20 µg/mL yielded increased cytotoxicity compared with controls [15]. As the concentration (>20 µg/mL) of SiNPs increased, the production of cytokines was correspondingly reduced [15]. Wang and coworkers revealed that, for poly-acrylic acid (PAA)-terminated SiNPs, undetectable in vitro cytotoxicity was observed with concentrations ranging from 0.1 to 200 µg/mL and incubation time from 24 to 48 h (Fig. 6.1) [22]. Recently, a comparative cytotoxicity study was performed to assess cytotoxicity of SiNPs with various surface functionalizations [27]. In this study, human colonic adenocarcinoma Caco-2 and rat alveolar macrophage NR8383 cells were used to clarify the toxicity of this series of SiNPs. Within the tested concentration range (0–100 µg/mL), the surface coating on the SiNPs appeared to dominate the cytotoxicity: the cationic SiNPs exhibited more cytotoxic than neutral SiNPs, whereas the carboxylic acid-terminated and hydrophilic PEG-or dextran-terminated SiNPs yielded feeble cytotoxicity [27]. Moreover, intracellular mitochondria seemed to be the target for the toxic cationic SiNPs since a dose-, surface charge- and size-dependent imbalance of the mitochondrial membrane potential was observed [27].

In addition, a series of other cellular events was resulted from cationic SiNPs, such as decrease mitochondrial membrane potential and ATP production, induction of ROS generation, increase of cytoplasmic Ca^{2+} content, production of TNF-α, and enhanced caspase-3 activity [27].

For clinical applications, it is essentially critical to investigate nanomaterials disposition and fate in the body. As a result, in addition to the above in vitro toxicity assessment, several pioneer studies have been carried out to investigate in vivo toxicity of the SiNPs. In 2011, Louie et al. evaluated the SiNPs biodistribution in mice by in vivo positron emission tomography (PET) [20]. In their study, amacrocyclic ligand-$^{64}Cu^{2+}$ complex were used to label dextran-coated manganese doped-SiNPs (HD: 15.1 ± 7.6 nm). The SiNPs were rapidly cleared from mouse blood stream and mainly accumulated in the liver, and finally excluded from body via real clearance [20]. In the following year, Galagudza et al. investigated the in vivo toxicity of SiNPs (mean diameter 10 nm) using both morphological and functional criteria [23]. They found that intravenous infusion of SiNPs showed little influence on hemodynamic parameters during long-term observation [23]. These results provide a demonstration of favorable biocompatibility of SiNPs in vivo. Recently, Ye et al. tested the toxicity assessment of Pluronic F127-encapsulated SiNPs (40–80 nm) with high intravenous doses (200 mg/kg) using mice and monkey as models [28]. Feeble toxicity in blood analysis was induced by the SiNPs during 3-month investigation. Of particular note, for the mice group, the SiNPs resulted in pathological changes in the liver. In contrary, no obvious change was observed toward the SiNPs-treated monkey under the same experiment conditions (e.g., dose and time). Levels of silicon were remarkably elevated in the liver, spleen, and kidneys of mice over 14-week period, which was to some extent inconsistent with the previously reported biodegradability of silicon in vivo [28].

Porous silicon nanoparticles (PSiNPs), recognized as another classic type of SiNPs, have been used in various applications including drug delivery, immunotherapy, and theranostics, etc. [30–32]. Notably, one typical property of porous silicon structures is their biodegradability in physiological environments, making them promising candidates for biological and biomedical applications [33, 34]. As a consequence, in the following part we give a brief discussion of the biocompatibility and biodegradability of PSiNPs.

In 2009, Sailor et al. revealed the PSiNPs with the mean diameter of 126 nm and pore diameter of 5–10 nm yielded feeble toxicity to HeLa cells within the tested concentration range (i.e., 0–0.2 mg ml^{-1}) by the calcein assay (Fig. 6.2a) [31]. For in vivo studies, the intravenously injected PSiNPs were mainly accumulated in the mononuclear phagocytic system (MPS)-related organs (e.g., liver and spleen), and then significantly or completely excluded from the body during 1 or 4 weeks, respectively (Fig. 6.2b). The possible reason was that the PSiNPs became water-soluble silicic acid via biodegradation. As a result, similar to the control mice, the body weight of the PSiNPs-treated mice slightly increased, providing additional demonstration of nontoxic properties of the PSiNPs (Fig. 6.2c). These results were also well confirmed by the haematoxylin and eosin

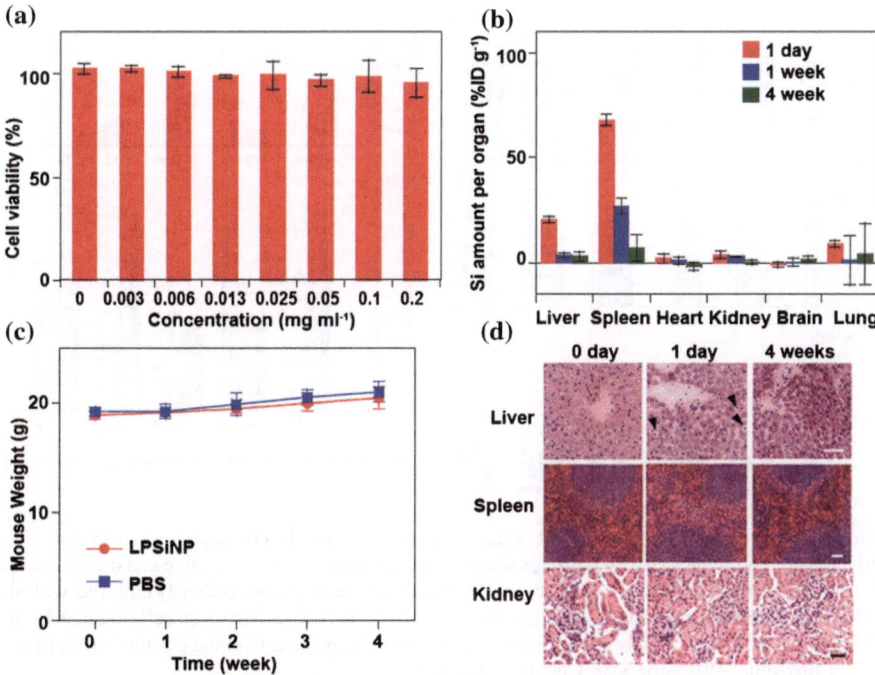

Fig. 6.2 a Cell viability of HeLa cells cultured with the pSiNPs with serial concentrations (0.003–0.2 mg/mL) for 48 h. **b** In vivo pSiNPs biodistribution in different dominant organs (e.g., liver, spleen, heart, kidney, brain, and lung) after 4-week intravenous injection of the pSiNPs (20 mg kg^{-1}). ICP-OES was employed for determining silicon concentration. **c** Quantitative measurement of body weight of mouse treated with the pSiNPs or PBS (as a control) for different time periods. **d** Histology of the liver, spleen, and kidney, collected from mice intravenously injected with pSiNPs (20 mg kg^{-1}). HE were used for staining organs. The PSiNPs swallowed by macrophages in the liver were indicated by the *arrows*. *Scale bars* stand for 50 μm. Reprinted by permission from Nature Publishing Group, a division of Macmillan Publishers Ltd. ref. [31], copyright 2009

(HE) assay, showing no inflammatory infiltrates in the liver and no obviously morphological change in the spleen and kidney after 4-week injection (Fig. 6.2d), indicating excellent biocompatibility of PSiNPs.

A relationship between toxicity and characteristic parameters of nanoparticles (e.g., chemical composition, size, shape, surface charge, surface chemistry, and so on) has been also investigated. Salonen and coworkers prepared thermally hydrocarbonized silicon (THCPSi) particles with several sizes (e.g., 97, 142, and 188 nm; 1–10, 10–25 μm) [35]. As shown in Fig. 6.3, the in vitro cytotoxic study showed dramatic decrease of cell viability, which was induced by the THCPSi particles with diameter of 1–10 μm. In comparison, the small size (142 nm in diameter) resulted in cellular damage and feeble cytotoxicity due to their relatively weak interaction with the cells. They further revealed that the Si particles produced different oxidative and cytotoxic responses to various cellular lines. In particular,

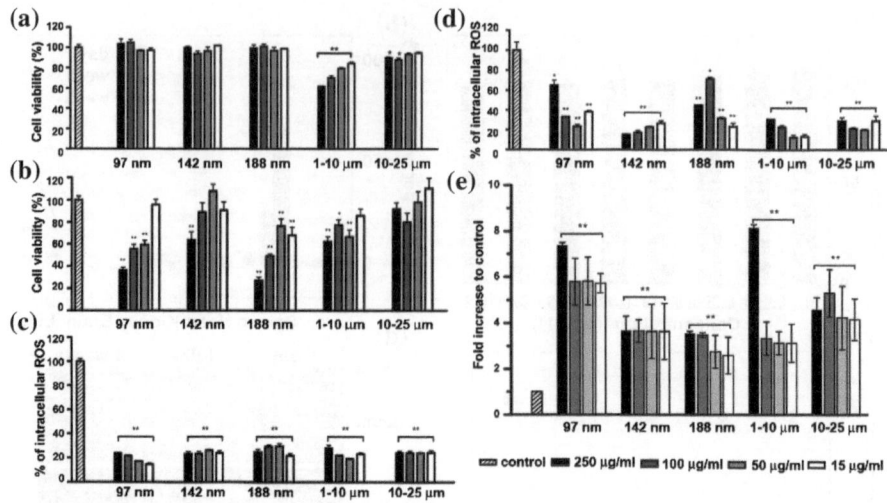

Fig. 6.3 **a** and **b** cell viability of Caco-2 cells (**a**) and RAW 264.7 macrophage cells (**b**) incubated with the THCPSi particles with various concentrations for 24 h. **c** and **d** Intracellular ROS measurement of Caco-2 cells (**c**) and RAW 264.7 macrophage cells (**d**) cultured with the THCPSi particles with various concentrations for 24 h. **e** Fold increase of cells treated by the THCPSi particles with different concentrations for 24 h. Reprinted with the permission from Ref. [35]. Copyright 2010 American Chemical Society

compared to human colon carcinoma (Caco-2) cells, RAW 264.7 macrophages displayed higher level oxidative and cytotoxic responses [35]. Similar results were reported by Bimbo et al., in which cell viability was largely dependent on the concentration and sizes of Si particles, as well as different types of cell lines [36].

The surface chemistry and surface charge also play important roles in the cytotoxic effect of PSiNPs, demonstrated by Santos and coworkers [37]. In their study, five different types of PSiNPs with similar size, surface area, and pore volume were prepared, namely TOPSi, thermally carbonized PSi (TCPSi) (3-Aminopropyl) triethoxysilane functionalized TCPSi (APSTCPSi), THCPSi, and undercylenic acid functionalized THPSi (UnTHCPSi). In general, PSiNPs cytotoxicity was more severely affected by the surface charge of the NPs than their hydrophilicity/hydrophobicity surface properties. For similar surface charges, PSiNPs with hydrophobic surfaces (e.g., THCPSi and UnTHCPSi) were more cytotoxic than the hydrophilic ones (e.g.,TOPSi and TCPSi). Among all the PSiNPs tested, the positively charge amine-modified PSiNPs (APSTCPSi) showed the highest extent of cytotoxicity via an indirect effect on the DNA. In contrast, negatively charged hydrophilic TOPSi and TCPSi NPs showed the lowest DNA damage effect [37]. The in vivo histological and biochemical studies also correlated rather well with the in vitro results [37]. The possible explanation for this observation was that the strong association of the positively charged amine-modified PSiNPs (APSTCPSi) with the negatively charged cell membranes could induce substantial damage of the cell membrane and even led to destruction [38]. In addition, the surface coating was regarded to be another contributor to influencing the in vivo biodistribution of PSiNPs. Sarparanta et al. [39] functionalized the THCPSi NPs with a self-assembled protein coating consisting of class II hydrophobin (HFBII, a kind of fungal protein from *Trichoderma reesei*) [40, 41]. They found that accumulation of the THCPSi NPs in the liver and spleen was significantly changed induced by biofunctionalization with HFBII (i.e., the liver-to-spleen ration of THCPSi uptake increased twofold after surface coating with HFBII) [35, 39].

6.2 Silicon Nanowires-Related Biosafety Assessment

SiNWs have received much research interests for various bioapplications to take advantage of their unique properties [42–48]. Therefore, systematic and reliable toxicity assessment of SiNWs becomes increasingly important, requiring thorough investigation before wide-ranging applications (e.g., fabrication of SiNWs-based biomedical devices).

Early in 2005, Coffer's group first demonstrated the ability of SiNWs, synthesized through vapor–liquid–solid (VLS) technique, to facilitate the growth of uniform, suggesting a noncytotoxic behavior of SiNWs [49]. Later, the same group evaluated the biocompatibility of SiNWs decorated with calcium phosphate (CaP/SiNWs) and the CaP/SiNWs composites modified with bisphosphonate (antiosteoporotic drug alendronate) [50]. Typically, the bisphosphonate-modified CaP/SiNWs exhibited more severe cytotoxicity than the CaP/SiNWs, suggesting surface functionality significantly influenced the overall biological response [50]. In 2007, Yang, Lee and coworkers employed HepG2 (a human hepatocellular carcinoma cell line) cells as models to investigate the cytotoxicity of SiNWs [51]. Treatment of cells with SiNWs suspensions (100 μg/mL) for 18 or 48 h led to change in the adhesion and spreading morphologies of HepG2 cells (Fig. 6.4). They further found that SiNWs with high concentrations yielded low cellular viability, indicating dose-dependent cytotoxicity of SiNWs [51]. Meanwhile, Yang et al. studied the interface of SiNWs array with mammalian cells, revealing the cells survived over 1 week despite the physical penetration of SiNWs [52]. Moreover, the longevity of the cells was found to be largely influenced by the diameter of SiNWs [52]. Thereafter, the same authors further studied the interactions between SiNWs arrays and cells [53]. They found that the SiNWs arrays had good biocompatibility (the cell viability was greater than 92 % within 48 h incubation time). On the other hand, compared to the flat surface, the SiNWs arrays enhanced the cell-substrate focal adhesion force and restrict cell spreading (the gene expression levels of Col I and α–actin were down-regulated) [53]. Notably, in later study, Piskin et al. showed that the SiNWs-treated cells exhibited normal morphology and high cell viability, suggesting feeble cytotoxicity of SiNWs [54].

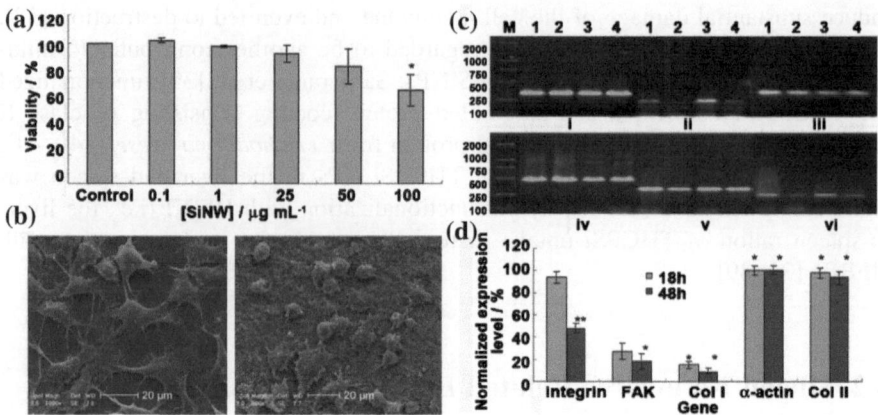

Fig. 6.4 **a** Cellular viability of HepG2 cells treated by the SiNWs with various concentrations ranging from 0.1 to100 μg/mL for 48 h at 37 °C, calculated by employing the Alamar blue assay. **b** SEM images of HepG2 cells cultured on a pure silicon wafer (*left*) or SiNW arrays growth on a silicon wafer (*right*). **c** Agarose gel (1.5 %) electrophoresis for RT-PCR products of adhesion-associated genes. (*i*), (*ii*), (*iii*), (*iv*), (*v*), and (*vi*) stand for β-actin, endogenous reference housekeeping gene, Col I, Col III, α-actin, integrin, and FAK, respectively. Marker in base pairs, control cells, SiNWs-treated cells (48 h), control cells, and SiNWs-treated cells (18 h) were categorized in lane 1, 2, 3, and 4, respectively. **d** Normalized expression level of adhesion-associated genes. QuantityOne (BioRad) was used for quantitative measurement of PCR products. Reproduced from Ref. [51] by permission of John Wiley & Sons Inc

The influence of topographical effects and chemical patterning of SiNWs arrays has been also investigated in recent years. In 2010, Yeh and coworkers developed nano-topographic oxidized silicon nanosphonges, which were suitable for modification with different chemicals, i.e., 3-aminopropyltrimethoxysilane (APTMS) with an amino group and perfluorodecyltrichlorosilane (FDTS) with fluorine [55]. Cell adhesion was found to be dominantly determined by the surface hydrophobicity. However, both the oxidized silicon substrates and nanosponges provided an environment for cell survival [55]. In 2011, Boukherroub et al. demonstrated that cells could be selectively cultured on superhydrophilic areas of patterned superhydrophobic/superhydrophilic SiNWs arrays [56]. They also showed that cell adhesion in the superhydrophilic regions was accompanied by SiNWs dissolution in the culture medium whereas the superhydrophobic surface remained unaffected [56]. Very recently, Robert and coworkers investigated the in vivo behavior of SiNWs (diameter: ∼20–30 nm, length: 2–15 μm) by using male Sprague-Dawley rats as models [57]. Typically, a dose-dependent increase in lung injury and inflammation was observed. Moreover, increase of lung collagen was detected after 91-day treatment of high-dose (e.g., 250 μg) SiNWs. The above results are not well consistent, which is possibly due to the different morphologies and surface properties of SiNWs used for the toxicity assessment.

6.3 Conclusions

In this chapter, we have discussed the biosafety assessment of silicon nanomaterials, by using the SiNPs and SiNWs as models. In terms of SiNPs, cytotoxicity is found to be greatly dependent on cell lines, properties of SiNPs (e.g., diameter, surface ligands, etc.), and incubation conditions (e.g., incubation time and SiNPs concentration). On the other hand, in vitro behaviors of SiNWs have been primarily studied, revealing that cellular viability well maintains when cells are incubated with SiNWs of relatively low concentrations (e.g., <10 µg mL^{-1}). However, high-concentration (>100 µg mL^{-1}) SiNWs produces obviously cytotoxicity to various cellular lines. In addition, length of SiNWs is regarded as an important parameter to influence cytotoxicity of SiNWs. It is worth noting that, in vivo experiment reveals that SiNPs may be biodegradable and readily be renally cleared from a mouse model with no detectable toxicity in vivo. However, other research claims that levels of silicon in liver, spleen, and kidney are detectable after 14-week post injection of mice. Such inconsistent results are possibly due to different properties of SiNPs (e.g., surface chemistry, component, morphology, etc.). Consequently, sufficient investigation is needed to clearly address this issue, which would be greatly valuable for silicon nanomaterials-based biological and biomedical applications.

References

1. Sharifi S, Behzadi S, Laurent S, Forrest ML, Stroeve P, Mahmoudi M (2012) Toxicity of nanomaterials. Chem Soc Rev 41(6):2323–2343
2. Elsaesser A, Howard CV (2012) Toxicology of nanoparticles. Adv Drug Delivery Rev 64(2):129–137
3. Long TC, Saleh N, Tilton RD, Lowry GV, Veronesi B (2006) Titanium dioxide (P25) produces reactive oxygen species in immortalized brain microglia (BV2): implications for nanoparticle neurotoxicity. Environ Sci Technol 40(14):4346–4352
4. Donaldson K, Tran L, Jimenez LA, Duffin R, Newby DE, Mills N, MacNee W, Stone V (2005) Combustion-derived nanoparticles: a review of their toxicology following inhalation exposure. Part Fibre Toxicol 2(1):10
5. Lewinski N, Colvin V, Drezek R (2008) Cytotoxicity of nanoparticles. Small 4(1):26–49
6. Fadeel B, Garcia-Bennett AE (2010) Better safe than sorry: understanding the toxicological properties of inorganic nanoparticles manufactured for biomedical applications. Adv Drug Delivery Rev 62(3):362–374
7. Aillon KL, Xie Y, El-Gendy N, Berkland CJ, Forrest ML (2009) Effects of nanomaterial physicochemical properties on in vivo toxicity. Adv Drug Delivery Rev 61(6):457–466
8. Lu YM, Zhong YL, Wang J, Su YY, Peng F, Zhou YF, Jiang XX, He Y (2013) Aqueous synthesized near-infrared-emitting quantum dots for RGD-based in vivo active tumour targeting. Nanotechnol 24 (13). doi:10.1088/0957-4484/24/13/135101
9. Winnik FM, Maysinger D (2013) Quantum dot cytotoxicity and ways to reduce it. Acc Chem Res 46(3):672–680
10. Andón FT, Fadeel B (2013) Programmed cell death: molecular mechanisms and implications for safety assessment of nanomaterials. Acc Chem Res 46(3):733–742

11. Boer WDAM de, Timmerman D, Dohnalova K, Yassievich IN, Zhang H, Buma WJ, Gregorkiewicz T (2010) Red spectral shift and enhanced quantum efficiency in phonon-free photoluminescence from silicon nanocrystals. Nat Nanotechnol 5 (12):878–884

12. Chrobak D, Tymiak N, Beaber A, Ugurlu O, Gerberich WW, Nowak R (2011) Deconfinement leads to changes in the nanoscale plasticity of silicon. Nat Nanotechnol 6(8):480–484

13. Erogbogbo F, Lin T, Tucciarone PM, LaJoie KM, Lai L, Patki GD, Prasad PN, Swihart MT (2013) On-demand hydrogen generation using nanosilicon: splitting water without light, heat, or electricity. Nano Lett 13(2):451–456

14. Alsharif NH, Berger CEM, Varanasi SS, Chao Y, Horrocks BR, Datta HK (2009) Alkyl-capped silicon nanocrystals lack cytotoxicity and have enhanced intracellular accumulation in malignant cells via cholesterol-dependent endocytosis. Small 5(2):221–228

15. Choi J, Zhang Q, Reipa V, Wang NS, Stratmeyer ME, Hitchins VM, Goering PL (2009) Comparison of cytotoxic and inflammatory responses of photoluminescent silicon nanoparticles with silicon micron-sized particles in RAW 264.7 macrophages. J Appl Toxicol 29(1):52–60

16. Shiohara A, Hanada S, Prabakar S, Fujioka K, Lim TH, Yamamoto K, Northcote PT, Tilley RD (2010) Chemical reactions on surface molecules attached to silicon quantum dots. J Am Chem Soc 132(1):248–253

17. Bhattacharjee S, de Haan LH, Evers NM, Jiang X, Marcelis AT, Zuilhof H, Rietjens IM, Alink GM (2010) Role of surface charge and oxidative stress in cytotoxicity of organic monolayer-coated silicon nanoparticles towards macrophage NR8383 cells. Part Fibre Toxicol 7(1):25

18. Erogbogbo F, Yong K-T, Roy I, Hu R, Law W-C, Zhao W, Ding H, Wu F, Kumar R, Swihart MT (2011) In vivo targeted cancer imaging, sentinel lymph node mapping and multi-channel imaging with biocompatible silicon nanocrystals. ACS Nano 5(1):413–423

19. Erogbogbo F, Tien C-A, Chang C-W, Yong K-T, Law W-C, Ding H, Roy I, Swihart MT, Prasad PN (2011) Bioconjugation of luminescent silicon quantum dots for selective uptake by cancer cells. Bioconjugate Chem 22(6):1081–1088

20. Tu C, Ma X, House A, Kauzlarich SM, Louie AY (2011) PET imaging and biodistribution of silicon quantum dots in mice. ACS Med Chem Lett 2(4):285–288

21. Sato K, Yokosuka S, Takigami Y, Hirakuri K, Fujioka K, Manome Y, Sukegawa H, Iwai H, Fukata N (2011) Size-tunable silicon/iron oxide hybrid nanoparticles with fluorescence, superparamagnetism, and biocompatibility. J Am Chem Soc 133(46):18626–18633

22. Wang Q, Bao Y, Zhang X, Coxon PR, Jayasooriya UA, Chao Y (2012) Uptake and toxicity studies of poly-acrylic acid functionalized silicon nanoparticles in cultured mammalian cells. Adv Healthcare Mater 1(2):189–198

23. Ivanov S, Zhuravsky S, Yukina G, Tomson V, Korolev D, Galagudza M (2012) In vivo toxicity of intravenously administered silica and silicon nanoparticles. Materials 5(10):1873–1889

24. Ohta S, Inasawa S, Yamaguchi Y (2012) Real time observation and kinetic modeling of the cellular uptake and removal of silicon quantum dots. Biomaterials 33(18):4639–4645

25. Ohta S, Shen P, Inasawa S, Yamaguchi Y (2012) Size- and surface chemistry-dependent intracellular localization of luminescent silicon quantum dot aggregates. J Mater Chem 22(21):10631–10638

26. Ahire JH, Wang Q, Coxon PR, Malhotra G, Brydson R, Chen R, Chao Y (2012) Highly luminescent and nontoxic amine-capped nanoparticles from porous silicon: synthesis and their use in biomedical imaging. ACS Appl Mater Interfaces 4(6):3285–3292

27. Bhattacharjee S, Rietjens IMCM, Singh MP, Atkins TM, Purkait TK, Xu Z, Regli S, Shukaliak A, Clark RJ, Mitchell BS, Alink GM, Marcelis ATM, Fink MJ, Veinot JGC, Kauzlarich SM, Zuilhof H (2013) Cytotoxicity of surface-functionalized silicon and germanium nanoparticles: the dominant role of surface charges. Nanoscale 5(11):4870–4883

28. Liu J, Erogbogbo F, Yong K-T, Ye L, Liu J, Hu R, Chen H, Hu Y, Yang Y, Yang J, Roy I, Karker NA, Swihart MT, Prasad PN (2013) Assessing clinical prospects of silicon quantum dots: studies in mice and monkeys. ACS Nano 7(8):7303–7310

29. Zhong Y, Peng F, Bao F, Wang S, Ji X, Yang L, Su Y, Lee S-T, He Y (2013) Large-scale aqueous synthesis of fluorescent and biocompatible silicon nanoparticles and their use as highly photostable biological probes. J Am Chem Soc 135(22):8350–8356

30. Anglin EJ, Cheng L, Freeman WR, Sailor MJ (2008) Porous silicon in drug delivery devices and materials. Adv Drug Delivery Rev 60(11):1266–1277

31. Park J-H, Gu L, von Maltzahn G, Ruoslahti E, Bhatia SN, Sailor MJ (2009) Biodegradable luminescent porous silicon nanoparticles for in vivo applications. Nat Mater 8(4):331–336

32. Gu L, Hall DJ, Qin Z, Anglin E, Joo J, Mooney DJ, Howell SB, Sailor MJ (2013) In vivo time-gated fluorescence imaging with biodegradable luminescent porous silicon nanoparticles. Nat Commun 4. doi:10.1038/ncomms3326

33. Canham LT (1995) Bioactive silicon structure fabrication through nanoetching techniques. Adv Mater 7(12):1033–1037

34. Canham LT, Reeves CL, Newey JP, Houlton MR, Cox TI, Buriak JM, Stewart MP (1999) Derivatized mesoporous silicon with dramatically improved stability in simulated human blood plasma. Adv Mater 11(18):1505–1507

35. Bimbo LM, Sarparanta M, Santos HA, Airaksinen AJ, Mäkilä E, Laaksonen T, Peltonen L, Lehto V-P, Hirvonen J, Salonen J (2010) Biocompatibility of thermally hydrocarbonized porous silicon nanoparticles and their biodistribution in rats. ACS Nano 4(6):3023–3032

36. Bimbo LM, Mäkilä E, Laaksonen T, Lehto V-P, Salonen J, Hirvonen J, Santos HA (2011) Drug permeation across intestinal epithelial cells using porous silicon nanoparticles. Biomaterials 32(10):2625–2633

37. Shahbazi M-A, Hamidi M, Mäkilä EM, Zhang H, Almeida PV, Kaasalainen M, Salonen JJ, Hirvonen JT, Santos HA (2013) The mechanisms of surface chemistry effects of mesoporous silicon nanoparticles on immunotoxicity and biocompatibility. Biomaterials 34(31):7776–7789

38. Tao Z, Toms BB, Goodisman J, Asefa T (2009) Mesoporosity and functional group dependent endocytosis and cytotoxicity of silica nanomaterials. Chem Res Toxicol 22(11):1869–1880

39. Sarparanta M, Bimbo LM, Rytkönen J, Mäkilä E, Laaksonen TJ, Laaksonen P, Nyman M, Salonen J, Linder MB, Hirvonen J, Santos HA, Airaksinen AJ (2012) Intravenous delivery of hydrophobin-functionalized porous silicon nanoparticles: stability, plasma protein adsorption and biodistribution. Mol Pharma 9(3):654–663

40. Hakanpää J, Paananen A, Askolin S, Nakari-Setälä T, Parkkinen T, Penttilä M, Linder MB, Rouvinen J (2004) Atomic resolution structure of the HFBII hydrophobin, a self-assembling amphiphile. J Biol Chem 279(1):534–539

41. Linder MB (2009) Hydrophobins: proteins that self assemble at interfaces. Curr Opin Colloid Interface Sci 14(5):356–363

42. Shao MW, Shan YY, Wong NB, Lee S-T (2005) Silicon nanowire sensors for bioanalytical applications: glucose and hydrogen peroxide detection. Adv Funct Mater 15(9):1478–1482

43. Patolsky F, Timko BP, Yu G, Fang Y, Greytak AB, Zheng G, Lieber CM (2006) Detection, stimulation, and inhibition of neuronal signals with high-density nanowire transistor arrays. Science 313(5790):1100–1104

44. Stern E, Vacic A, Rajan NK, Criscione JM, Park J, Ilic BR, Mooney DJ, Reed MA, Fahmy TM (2010) Label-free biomarker detection from whole blood. Nat Nanotechnol 5(2):138–142

45. Wei X, Su S, Guo Y, Jiang X, Zhong Y, Su Y, Fan CH, Lee S-T, He Y (2013) A molecular beacon-based signal-off surface-enhanced raman scattering strategy for highly sensitive, reproducible, and multiplexed DNA detection. Small 9(15):2493–2499

46. Schmidt V, Wittemann JV, Senz S, Gösele U (2009) Silicon nanowires: a review on aspects of their growth and their electrical properties. Adv Mater 21(25–26):2681–2702

47. Su Y, Wei X, Peng F, Zhong Y, Lu Y, Su S, Xu T, Lee S-T, He Y (2012) Gold nanoparticles-decorated silicon nanowires as highly efficient near-infrared hyperthermia agents for cancer cells destruction. Nano Lett 12(4):1845–1850
48. Peng K-Q, Wang X, Li L, Hu Y, Lee S-T (2013) Silicon nanowires for advanced energy conversion and storage. Nano Today 8(1):75–97
49. Nagesha DK, Whitehead MA, Coffer JL (2005) Biorelevant calcification and non-cytotoxic behavior in silicon nanowires. Adv Mater 17(7):921–924
50. Jiang K, Fan D, Belabassi Y, Akkaraju G, Montchamp J-L, Coffer JL (2009) Medicinal surface modification of silicon nanowires: impact on calcification and stromal cell proliferation. ACS Appl Mater Interfaces 1(2):266–269
51. Qi S, Yi C, Chen W, Fong CC, Lee S-T, Yang M (2007) Effects of silicon nanowires on HepG2 cell adhesion and spreading. Chem Bio Chem 8(10):1115–1118
52. Kim W, Ng JK, Kunitake ME, Conklin BR, Yang P (2007) Interfacing silicon nanowires with mammalian cells. J Am Chem Soc 129(23):7228–7229
53. Qi S, Yi C, Ji S, Fong C-C, Yang M (2009) Cell adhesion and spreading behavior on vertically aligned silicon nanowire arrays. ACS Appl Mater Interfaces 1(1):30–34
54. Garipcan B, Odabas S, Demirel G, Burger J, Nonnenmann SS, Coster MT, Gallo EM, Nabet B, Spanier JE, Piskin E (2011) In vitro biocompatibility of n-type and undoped silicon nanowires. Adv Eng Mater 13(1–2):B3–B9
55. Yang C-Y, Huang L-Y, Shen T-L, Yeh JA (2010) Cell adhesion, morphology and biochemistry on nano-topographic oxidized silicon surfaces. Eur Cell Mater 20:415–430
56. Piret G, Galopin E, Coffinier Y, Boukherroub R, Legrand D, Slomianny C (2011) Culture of mammalian cells on patterned superhydrophilic/superhydrophobic silicon nanowire arrays. Soft Matter 7(18):8642–8649
57. Roberts JR, Mercer RR, Chapman RS, Cohen GM, Bangsaruntip S, Schwegler-Berry D, Scabilloni JF, Castranova V, Antonini JM, Leonard SS (2012) Pulmonary toxicity, distribution, and clearance of intratracheally instilled silicon nanowires in rats. J Nanomater 2012:17

Chapter 7
Outlook

7.1 Conclusion

In this chapter, we summarized representative and promising achievement to highlight the remarkable development of silicon nanotechnology for biological and biomedical applications in recent years, with the hope to promote the awareness of the realm of silicon nano-biotechnology. The past two decades have witnessed vast advancement of designing and fabricating a large number of silicon nanomaterials with well-defined structures and desirable functionalities. In particular, a variety of fabrication techniques have been developed, allowing rational design of low-dimensional silicon nanostructure and silicon-based nanohybrids. To date, fluorescent silicon nanoparticles (SiNPs), as the most important zero-dimensional silicon nanostructures, can be readily achieved *via* "top-down" or "bottom-up" methods using large-size silicon materials (e.g., silicon wafers used in electro-chemical etching methods) or organometallic silicon molecules (e.g., 3-(amino-propyl) trimethoxysilane employed in microwave-assisted strategies) as reaction precursors, respectively. Meanwhile, a number of approaches have been explored for synthesizing silicon nanowires (SiNWs), among which chemical vapor deposition (CVD), oxide-assisted growth (OAG), and metal-catalyzed electroless etching approach are recognized as well-established means capable of large-scale preparation of SiNWs with high production yield. In addition, multifunctional silicon-based nanohybrids (e.g., metal NPs-decorated SiNWs and metal element-doped SiNPs) have been rationally designed with desirable features. These kinds of well-developed silicon-based nanomaterials and nanofabrication techniques impart significant momentum to the development of silicon nano-biotechnology, opening up exciting avenues in biological and biomedical applications, particularly in biosensing, bioimaging, and cancer therapy.

Silicon nanomaterials featuring unique electronic/optical/mechanical properties have been widely employed for constructing a number of biosensing devices with excellent sensitivity and specificity, high reproducibility, and multiplexing capabilities. SiNWs-based field-effect transistor (FET), as one representative kind of silicon-based electrochemical biosensor, has drawn intensive investigation during

Y. He and Y. Su, *Silicon Nano-biotechnology*, SpringerBriefs in Molecular Science, DOI: 10.1007/978-3-642-54668-6_7, © The Author(s) 2014

the past decade. A variety of biological species (e.g., nucleic acids, proteins, and virus, etc.) can be readily and sensitively detected using the SiNWs-based FET devices. On the other hand, silicon-based hybrids (e.g., AgNPs-decorated SiNWs, AuNPs-coated silicon wafers) have been widely explored as highly active SERS substrates, and further been employed for construction of high-performance SERS biosensors, enabling specific and reproducible detection of DNA and proteins with extremely low concentrations, even down to $\sim$ fM levels. Notably, recent reports revealed that the silicon-based SERS nanosensors, serving as novel in vitro sensing platform, may afford new opportunities for detecting cells and analyzing cellular behaviors at the single-cell level.

Proof-of-concept studies have opened up promising avenues for establishing novel silicon nanotechnology-based bioimaging techniques to take advantages of high-performance silicon-based nanoprobes. Compared to well-established fluorescent proteins/organic dyes-based bioprobes of severe photobleaching property and II-VI fluorescent QDs-based nanoprobes involving heavy metal-induced safety concerns, SiNPs feature strong fluorescence, robust photostability, and excellent biocompatibility. Those attractive merits have triggered extensive exploration of SiNPs as potentially ideal biological fluorescent probes. Many reported papers have provided sufficient demonstration that SiNPs-based bioprobes with superior optical properties are particularly suitable for long-term and real-time bioimaging applications. Moreover, multifunctional silicon-based nanoprobes have been utilized for multimode bioimaging studies. Specifically, fluorescent and magnetic SiNPs-based probes have been demonstrated to be ready for simultaneous fluorescence and magnetic resonance imaging (MRI) in vitro and in vivo in a high-contrast manner.

Silicon-based nanoagents have appeared as novel therapeutic agents for treatment of cancer with encouraging outcomes. Plenty of work has focused on developing porous silicon (pSi) as high-quality delivery systems for drug molecules, therapeutic proteins and genes, facilitating improvement of treatment effect and decrease of toxic adverse effects. Recent investigation reveals SiNWs as promising drug nanocarriers featuring ultrahigh drug-loading capacity of $\sim$ 20,800 mg/g, significantly larger than the values reported for other kinds of nanomaterials-based drug carriers. In vivo experiment demonstrates the SiNWs-based drug nanocarriers are highly efficacious for long-term inhibition of tumor growth in a mice model. In addition to chemotherapy, the silicon-based nanoagents show a possibility for phototherapy of cancer. Typically, pSi-based photosensitizers are capable of photodynamic therapy (PDT) for efficiently destroying various malignant cell lines. SiNWs-based hyperthermia nanoagents are highly efficacious for photothermal therapy (PTT), leading to complete destruction of cancer cells under specific light irradiation.

To bring silicon nanotechnology for wide-ranging applications, reliable and comprehensive biosafety assessment should be first carried out, while silicon is creditable for non or low toxicity. By using SiNPs and SiNWs as models, pioneering work has suggested favorable biocompatibility of silicon nanomaterials, i.e., feeble toxicity in vitro and in vivo is observed when cell and mice are treated with SiNPs or SiNWs. Notwithstanding, some primary research expresses silicon nanomaterials-induced safety concerns, claiming that silicon nanomaterials would lead to detectable adverse effects on cell viability when concentrations of silicon nanomaterials reach a threshold or their surface is modified by special kinds of ligands.

7.2 Challenges and Perspectives

While silicon nanotechnology has shown great promise for myriad biological and biomedical applications, there still remain a number of central challenges facing rational design of well-defined silicon nanostructures, sophisticated fabrication of silicon-based sensing devices, nanoprobes, and nanoagents, as well as reliable and systematic risk assessment.

First, despite tremendous progresses of synthesis of silicon nanomaterials, extensive efforts are still required to address following critical issues, including: (1) Exact fluorescent mechanism of SiNPs is required to be verified, which would provide valuable theoretical guidance for rationally fabricating fluorescent SiNPs with tunable optical properties. (2) Scientists should put continuous focus on controllable and uniform growth of SiNWs of desirable diameters and lengths. (3) Development of facile, rapid, and low-cost synthetic methods for large-scale preparation of silicon nanomaterials is of essential importance for practical applications of silicon nanotechnology. Moreover, based on previously reported results and our latest experiment results, we envision that surface ligands may probably serve as a significant contributor to chemical/physical properties of SiNPs. However, sufficient theoretical and experimental data are urgently required to clearly clarify the relationship between interface and properties of silicon nanomaterials, which would vastly facilitate rational design and fabrication of silicon nanomaterials with well-defined structures and required functionalities, further providing a solid guarantee for silicon nanotechnology-based biological and biomedical applications.

Second, while a number of silicon-based sensing devices have been constructed for high-sensitivity and high-specificity bioassays, satisfactory sensing mechanisms are unclear, which is adverse to further optimizing performance (e.g., sensitivity, specificity, and reproducibility) of the sensors. It is worthwhile to point out that, most of the reported detection cases are limited in experimental environment thus far. Therefore, tremendous efforts are required to employ the silicon-based sensors for detection of biological species in real systems, providing a consolidate feasibility evaluation for practical applications. Moreover, fabrication of cheap silicon-based sensors in a low-cost manner is another critical issue for widespread sensing applications.

Fig. 7.1 Reports on silicon nanotechnology for biological and biomedical applications are highlighted as covers of Angew. Chem. Int. Ed., indicating silicon nano-biotechnology may serve as highly promising platform for myriad applications, including biosensing, bioimaging, and cancer therapy, etc. Reproduced with permission of John Wiley & Sons Inc

Third, for silicon nanotechnology-based bioimaging applications, most reported publication focuses on the use of SiNPs-based nanoprobes for single-color cellular imaging. In contrary, multicolor bioimaging is more suitable for providing abundant information of cells and tissues, which would be considered as an important further research direction. Moreover, while several reports have successfully demonstrated feasibility of long-term bioimaging, fixed cells are generally used as models in these cases. Scientists should make intensive efforts to realize real-time and long-term tracking of live cell using the high-quality silicon-based nanoprobes, which is much more scientifically significant. In addition, compared to sufficient reports on in vitro imaging, there exists relatively scanty information on silicon-based bioprobes for bioimaging in animals. Consequently, based on the development of novel multifunctional silicon-based nanoprobes, extensive efforts should be devoted to employing silicon nanotechnology for in vivo imaging applications.

Fourth, in terms of cancer therapy, although several silicon-based nanoagents have been reported to date, they are merely ready for single-model treatment of cancer. Multimodel therapeutic strategies are recently found to be highly efficient for cancer diagnosis and therapy. As a result, we should further develop more kinds of silicon-based multifunctional nanoagents to meet increasing demand of cancer treatment. More importantly, to improve therapeutic efficiency and reduce toxic side effects, exact in vitro and in vivo behaviors of the silicon-based nanoagents should be investigated in a detailed way, which is nevertheless remaining unknown to date.

Last but not the least, systematic risk assessment of silicon nanotechnology should be comprehensively carried out in a reliable way, providing a feasibility evaluation of using silicon nano-biotechnology for widespread application.

Provided these mentioned issues are satisfactorily addressed, silicon nano-bio-technology may be anticipated to serve as a powerful platform for remarkably facilitating advancement of scientific research and clinic applications in the future (Fig. 7.1).